FLYING THE
WEATHER MAP

Also by
Richard L. Collins

FLYING SAFELY

FLYING IFR

FLYING THE WEATHER MAP

Richard L. Collins

Delacorte Press/Eleanor Friede

Published by
Delacorte Press/Eleanor Friede
1 Dag Hammarskjold Plaza
New York, N.Y. 10017

Manufactured in the United States of America

First printing

Designed by MaryJane DiMassi

LIBRARY OF CONGRESS CATALOGING IN PUBLICATION DATA

Collins, Richard L   1933–
  Flying the weather map.

"An Eleanor Friede book."
  Includes index.
  1. Meteorology in aeronautics.   I. Title.
TL556.C58   629.132′4   79–10916
ISBN 0–440–02610–5

Contents

Foreword

There is no question that weather is of primary importance to pilots. When using an airplane for cross-country flying, a basic understanding of the elements can mean the difference between success and failure. And learning about weather, unlike learning about many other things, is not just a matter of black/white relationships. Weather is a soft, pliable thing that slips around, in, and through the minds of pilots, meteorologists, and computers. Forecasts are not totally accurate and, unlike fine drink, they worsen with age. The actual weather reports that we get are accurate enough, but they represent conditions only at selected spots. And while weather maps are usually accurate, they represent history, a weather situation that existed in the past. Perhaps it was just an hour ago, but it might have been several hours ago. Meterology texts (and this isn't one) tend not to relate to the real flying world.

That is not to paint a hopeless, or even a difficult picture. Aviation weather is both understandable and logical. A pilot needs only an understanding of the basics, some experience, and an open mind. The last two items are especially important, because experience teaches us that while we have to examine black/white weather information, when flying we spend most of our time squinting at various shades of gray. Time helps us

learn to interpret the shades. The open mind is required because what we see and experience can be at odds to what we
thought we would see and experience, and it is often necessary
to go back to square one and begin anew to try and understand
the elements as they are affecting a flight. This understanding
can then be stored for future reference.

This book starts with an exploration of some basics of
weather. It is not a textbook, it is a discussion of the staples,
with emphasis on the things that affect light airplane flying.
This is followed by accounts of actual cross-country flights at
all times of the year. In each case, the information received
before takeoff is compared with actual weather conditions and
with the basics of meteorology. Most flights were conducted
IFR, but the relationship to VFR flying is also explored. The
other necessary ingredient, the open mind, is yours to provide.

—Richard L. Collins

FLYING THE WEATHER MAP

Low Pressure

If flying the weather was a game with only one specific question allowed before each flight, what would be the best thing to ask? There is no doubt that potentially the most useful question would be: "Where are the low pressure centers?"

Lows are weather makers. They can affect wide areas, and if we don't understand anything else about meteorology, we had better understand some basics about low pressure. When all the forecasts seem in error, and there are a lot more questions than answers, a low is probably misbehaving. It is moving faster or slower than anticipated, or perhaps it is deepening (strengthening) or filling (becoming weaker). Or maybe a low formed in an unexpected location. Only if we know the characteristics of a low can we construct a mental image of weather situations—especially of situations that are changing rapidly and defying the efforts of forecasters.

The basic circulation around a low is counterclockwise in the northern hemisphere, which will be the area of discussion throughout the book. Air also flows into a low. If it flows into a low and stops, the low begins to fill and the area of low pressure will simply go away. For the low to live on, air must flow into the low and then move upward in the atmosphere.

We must understand that upper air patterns have a lot to do

with surface systems. The most important key to upper level influence is found on the 500 millibar chart. This reflects the pressure patterns at approximately 18,000 feet, which is the center of the vertical distribution of mass. There are charts for higher levels, but there is little change in the patterns from the 500 millibar surface on up. We will discuss the relationship of the 500 millibar chart to surface weather in Chapter 3, and use it extensively in examining weather situations in the last four chapters of the book.

MOISTURE

Moisture supply has a lot to do with weather, and if we compare geography with the counterclockwise circulation around low pressure centers, we get a good idea of how this works.

When a low is to the west of a given point, the circulation at that location will be from the south. That means relatively warm air, and potentially moist air, depending on the area from which the air is being drawn. Basic meteorology teaches us that air assumes the properties of the surface over which it flows, and the effects of southerly circulation at different locations clearly illustrates this.

Start at Denver, where a south wind comes from over old and New Mexico. This is dry country, and Denver's average annual precipitation totals only fifteen inches a year.

Move now to Wichita, 400 miles to the east. There is more potential circulation from over the Gulf of Mexico and the average annual precipitation moves up to twenty-eight inches.

Move even farther east, to Little Rock, directly north of the Gulf, and note an average annual precipitation of forty-eight inches. This forty-eight inches, incidentally, is about an average for cities that are directly north of the Gulf of Mexico and inland a few hundred miles. Montgomery, Alabama, and Atlanta both have about the same average annual precipitation as Little Rock.

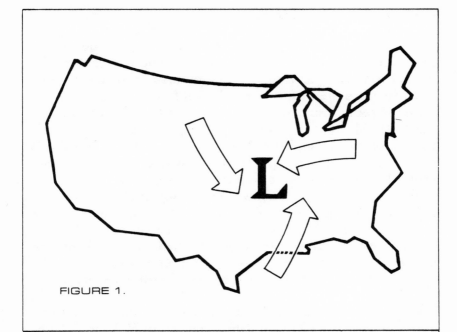

FIGURE 1.

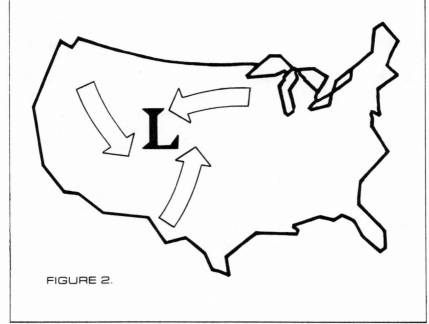

FIGURE 2.

AROUND THE LOW

Put a low pressure center on a map and visualize the circulation around it. Figure 1 is an example. The low center over central Arkansas has three circulation arrows for illustration. Relate each to the surface over which it flows. The one coming from the south brings warm and moist air to the low from over the Gulf. This is the basic feed of the system. The easterly circulation brings moisture from over the Atlantic but this is not as warm and perhaps not as wet as that southerly arrow. From the northwest we have dry and cold air.

For contrast, consider a low over the northwestern corner of Colorado, as in Figure 2. Not one bit of circulation has close access to a lot of moisture. The low in Colorado might be characterized more by strong winds than by frog-strangling precipitation. Indeed, there is a sign in Gunnison, Colorado, proclaiming that the sun shines at least a little on every day of the year.

Low pressure areas don't often remain stationary, but when one does, the rising air is mostly in the center of the low. A low that remains stationary can become quite strong, with increasing circulation as it deepens. Everything eventually has to move, though, and when a low pressure area is on its way, the major ascent of air in the low is usually on the forward side, in the direction of motion, which is usually to the east or the northeast. Thus we would usually find clouds stacked high ahead of a moving low as warm, moist air moving up from the south at low levels is lifted over a wide area. And, a moving low can maintain strength as it feeds on the warm moist air moving northward ahead of it.

In seeking information about weather, we occasionally hear people remark that there is "no weather" in connection with a given low pressure system. This might be true today, but things can sure change by tomorrow. The low in Colorado is dry because it doesn't have a good moisture source from which to brew clouds and precipitation. Move it 800 miles to the east and the character or the system might change completely.

UPPER AIR PATTERNS

When we acknowledge that air rises in a low, we see the basis for the effect of upper air pressure patterns and circulations on the severity of low pressure systems. The upper patterns contribute to the development of lows, and steer them once they have formed. Some very specific effects will be covered in Chapter 3. For now, just remember that the weather map on the wall at the flight service station, and the one shown on TV, is a surface weather map. If a low is depicted with little weather around it, that is because of geography or lack of support in the upper air circulation. As the low moves, the geography around it changes. As time passes, the surface low's relationship with upper air patterns changes. The primary point is that no low pressure system should be discounted until it is well east of your position. If it's still coming your way, it can change and change rapidly. Or if conditions to the west, southwest, or south are ripe for low pressure development, a clear day can turn cruddy.

BIRTH OF A LOW

The development of a low pressure system is worth examining briefly here.

Basically, a low can be formed when dissimilar air masses rub against each other. Think of the breakers and the undertow at the beach. Water is moving toward the shore, and it is also moving away from the shore. Friction develops between the two (and/or between water moving toward the shore and the bottom) and, presto, the water curls up and over into a breaking wave. Turning to weather, if you have a southwesterly circulation next to a northeasterly circulation, the same thing can happen. A wave can develop and it can curl up and over and form a low pressure center. Sometimes the circulation does not loop quite all the way over and a full circulation doesn't form; instead, waves develop and move along the boundary between the two air masses. These are called low pressure waves.

There is a strong relationship between low pressure systems

and temperature. We all know about fronts that develop around lows, and we will talk about these in Chapter 2. For now, just think of there being differences in temperature around a low. Warm air is going to be ahead of the low, cold air behind it. Warm air has the capability of holding far more moisture than cold air and, in fact, large amounts of moisture can be moved only in warm air. This brings the low what it needs; the upward swirl results in the cooling of the warm air which creates clouds and precipitation.

The temperature differences support the development and life of a low pressure system. A low tends to be most active when working with great temperature differences—witness the tornado outbreaks in the springtime—and late in the life of a low its circulation mixes things up so there's not so much difference in temperature. Then the low starts to weaken, to dissipate, done in by the work of its circulation.

What is the ideal condition for low development? One classic pattern is found in the southern part of the U. S. in the fall, winter, or spring, when the circulation around an old low far to the northeast pushes cold air down close to the Gulf of Mexico. There it stops for lack of enough push from that far-away low to keep moving. This leaves a situation with cold air to the north and warm air to the south; northeast wind next to southwest wind. It's a situation ripe for development.

PRESSURE

How low does the pressure have to be for a low to make real trouble? That depends on a lot of things. The difference in pressure between the low and the nearest high, and the distance involved in this pressure change, determines the strength of the circulation. The moisture supply is a big factor. There are other factors, but circulation is an obvious thing to look for in seeking clues to a low's effect on flying weather. Surface wind and wind aloft are good indicators of low strength, although the weather may turn bad for VFR flying before the surface wind suggests the approach of a low. So add to wind the actual weather that is developing in the area between you and the low, as a guide to what might be expected.

Also consider that where the temperature of a large body of water is warmer than the surface temperature, as is often true in the winter, the temperature and pressure differences of air masses can be accentuated in the low levels, and deep low pressure systems can form quickly over the open water.

HIGH PRESSURE

High pressure is the opposite of low pressure in many respects. A high tends to be colder, a low warmer; a high drier, a low wetter. The flow around a high is clockwise and outward. In studying aviation weather, the location of highs is not as critical as the location of lows, although there are times when a high pressure area might make weather. One example is when a high is situated north of a given location on the east coast. This causes an easterly onshore flow, and warm moist air flowing over cold ground can result in low cloudiness. Too, the location of the high pressure centers is at times useful in estimating and minimizing the effects of the wind aloft. Flying toward the center of a high, going from east to west, we'll first see northwesterly, then light, then southerly winds, as in Figure 3.

In discussing the movement of air into a low, and its ascent, we moved a lot of air upward. It can't all stay up, and a high being the opposite of a low, we'll find descending air in the center of the high. If air could be tagged like pigeons, we might even find a parcel of air that moved up from the south and into the low where it ascended, lost all its moisture, and then wandered over above a high and came back for another look at the surface.

If we just remember the part about air flowing into a low, and coming back down in the middle of a high, we can harbor a basic understanding of the weather created by each in most situations. The air comes to the low laden with moisture. In it goes and, as the air ascends, the moisture falls out of it, in the form of rain or snow or whatever. All that moisture is gone as the air settles back in the middle of a high, and it brings the clear and cold air from aloft down upon us. That's rather simplistic and there are variations, but perhaps the generaliza-

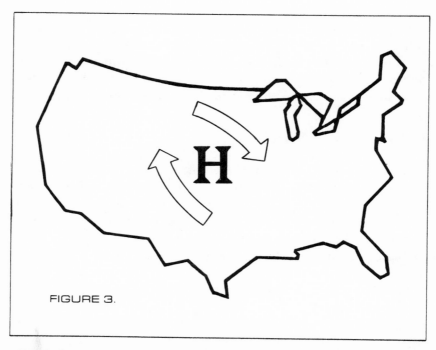

FIGURE 3.

tion helps. Combine it with the direction of circulation (counterclockwise around a low, opposite around a high) and you
can build a weather map in your head when knowing only the
location of the pressure systems, especially the low pressure
systems. Or you can get some idea of what is going on just by
knowing the wind direction and velocity at your position.

SIMPLE RULES

The counterclockwise flow around low pressure centers also
leaves us with a nice and simple rule-of-thumb to use in con-
templating weather while aloft. If flying with a tailwind, better
weather is more likely to be found to the right. No matter
where you are around the low, if the wind is from behind, a
right turn will take you away from the low center. If the wind
is from ahead, better weather will be to the left. Another basic
rule tells us that a low center is probably stronger than anticipated if the winds are stronger than forecast.

Temperature, Fronts, and Troughs

Temperature is a key product of weather as well as a key consideration in flying. A good example of this came at the time I was writing the first draft of this chapter. In the evening, the forecaster said the next day would be really bad in New Jersey, with a chance of heavy snow in the area. It certainly didn't sound like it would be much of a day for flying. When I got up the next morning, though, the temperature was +10C. The forecaster was correct about precipitation—it was pouring rain —but, fortunately, he missed on temperature. This meant that the low pressure center did not develop or move as expected. The surface flow over the area was southeasterly to southerly around a low to the west, instead of easterly to northeasterly around a low to the south that they thought might form off the Carolina coast. It didn't form and the day would have been perfectly fit for IFR flying without deicing equipment.

Moving air of various temperature around is what weather is all about, anyway. If there is no change in temperature, much change in weather is not likely. The relationship of temperature to the weather map is plain to see in Figure 4. You can even tell about where the low is just by applying the basics of circulation to the temperature patterns.

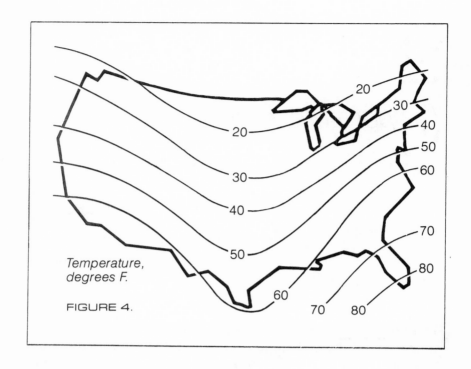

Temperature,
degrees F.

FIGURE 4.

RULE-OF-THUMB

Some of the basics of highs and lows intermingle with temperature to expand rules-of-thumb. If the weather is much colder than normal for the time of year, the flying weather might be VFR more often than usual. For example, the record-setting cold winter of 1976–77 was a pretty good time to fly—if you could get the airplane started. If the weather is warmer than normal, the flying weather is more likely to be IFR, especially in the fall, winter, or spring. A good example was November 1977. Except for a cold snap or two, it was rather mild over the eastern part of the country. It was also very wet—twelve or more inches of rain fell in areas that normally have only a few inches in that month. And the flying was very IFR. I flew sixty-seven hours that month in my single-engine Cessna, and twenty-four of the sixty-seven hours were flown in cloud. It was the most IFR month that I can remember.

Cold equates with surface high pressure, warm with low

pressure, the latter especially if we consider the area most affected by the low—that territory to the east of the low center. This relationship tells us why the bad weather comes with the approach of a low. Warm air will hold more moisture than cold air; in fact, for every 11 degrees C rise in temperature, the capacity of air to hold water vapor is about doubled. Saturate that nice warm air near the surface and ahead of a low, then draw it toward the low and lift and cool it, reducing its capacity to hold water, and presto, rain or snow.

DEWPOINT

Weather sequences include both the temperature and the dewpoint. The dewpoint number signifies the level to which the temperature has to drop for the air to become saturated, creating conditions conducive to condensation and the formation of fog or clouds. Every pilot knows that when the temperature and dewpoint are together, or very close, the weather is probably pretty bad, or at least suspect.

The temperature can match the dewpoint because of several factors. At night, we watch for ground fog as the surface temperature drops and the dewpoint remains the same. Or the dewpoint can increase as the temperature remains the same if water vapor is added to the air, as when a warm rain falls into colder air (or when you take a hot shower in a cold bathroom). Temperature dropping and/or dewpoint increasing is important because fog can cause an airport to go below minimum quickly. Warm rain falling into colder air can cause low scud to form, and that scud can become scattered, then broken, and perhaps even overcast. The next time it starts raining from a high overcast, watch for the formation of low clouds. We often think of clouds as moving in from somewhere, but they can just form, because of precipitation adding water vapor and raising the dewpoint.

Another way the temperature and dewpoint get together is when air is lifted. When this happens, a reduction in pressure takes place and the air expands and cools. When it cools to the dewpoint, condensation occurs and, given any moisture in the air, clouds form. Lifting can be caused by differential heating of the earth, as when summertime cumulus is formed; or by wind flowing over rough terrain, forming the row of cumulus that perches along ridges; or by vertical air motion, as air swirls around and into a low pressure system.

The rate at which air cools with altitude has a direct relationship on the action. If it is lifted and cooled very slowly, or if warm air overlays cold, the air is stable. Not much going up, not much going on. If the air is lifted and cooled rapidly, it is considered unstable. The reduction in pressure with altitude is rapid and the ability to retain moisture is decreased rapidly. There is a lot of action and the upward movement can continue.

As far as temperature goes, a drop of more than 2 degrees C per 1,000 feet is an indication of instability. That 2 degrees is the rate at which moist air naturally cools. A general temperature drop greater than that means that moist air will be moving upward into air that is cooling more rapidly than it will cool. Warmer air in colder surroundings continues to rise, even to accelerate, and it's of such stuff that thunderstorms are made.

When looking at a cross-section of the atmosphere, the temperature does not always exhibit an even decline with altitude. For example, warm air might move over colder air along a warm frontal slope, resulting in a temperature increase with altitude over a given point. But in this case, if you could take the temperature of a parcel of air and continue to measure it as the parcel moved northward, over the slope of the front, it would decrease continuously with altitude above the slope. Concern about temperature must cover all altitudes, not just those near the surface.

CIRCULATION AND TEMPERATURE

The circulation around low pressure systems results in the mixing of air of various temperatures and moisture contents. If the dissimilar air around a low is mixed rapidly—if the circulation is reasonably strong—lines of demarcation between warm and cold air develop. These are our basic fronts, cold and warm, named for the temperature of the advancing air relative to the retreating air.

Figure 5 is a drawing of the classic low pressure system and the circulation around it. The cold front is trailing to the southwest of the low and the warm front is off to the east. It looks a bit like a pie cut into one large and one small piece, but remember that this is a surface map. The actual effect of the interacting cold and warm air is much more equal than suggested. Warm air flows up over cold air, and the effect of the northward moving warm air is noted aloft far to the north of the surface warm front. The cloud system of a warm front actually extends 600 or 700 miles ahead of the surface position of the front. The easterly arrow north of the warm front represents surface flow *only*. Aloft, the wind would still be southerly or southwesterly.

WARM FRONT

A warm front has a shallow slope, as the lifting is gradual. Because the air is warm and moisture-laden, condensation and cloud formation occur as it moves up over the cold air. The

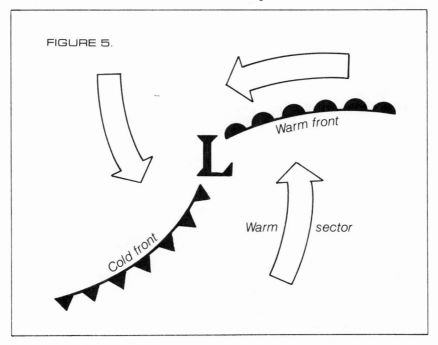

FIGURE 5.

weather in the warm sector, that area south of the warm front, is often relatively good until you reach the surface front, where the lifting begins. At that point, conditions might begin changing from reasonably good but perhaps showery weather to generally low ceilings and poor visibilities.

As pilots we learn that the weather north of the surface position of a warm front is characterized by cruddy conditions over a large area. The precipitation is caused by lifting and cooling, condensation and saturation. Warm rain falling into cool air below can raise the dewpoint at low levels to the temperature level, and low clouds might form below the slope of the warm front.

Consider two sets of clouds ahead of a warm front: those formed by lifting along a frontal slope, and those formed as rain falls into cooler air. The closer you are to the front, the lower the frontal slope, and the more likely the two sets of clouds will merge.

The VFR pilot is usually shot down by a warm front because of low ceilings ahead of the front, where the warmer rain is falling into cooler air. Usually these low ceilings are not obstructive to the IFR pilot—they don't often lower conditions below instrument landing system approach minimums—but they can create large areas of minimum weather, making the selection of an alternate difficult unless you happen to be flying awash with fuel.

Thunderstorms embedded in other clouds are quite another matter for the IFR pilot. They are often forecast in connection with warm fronts, and the mention of them is enough to make a pilot without weather avoidance gear wonder whether or not it is wise to fly.

A lot of factors bear on the formation of storms in a warm frontal zone. The amount of moisture, the strength of the circulation around the low, and the vertical temperature distribution all bear on the question. A slight plus is that storms that do develop in a warm frontal zone tend not to be as severe as storms that develop ahead of a cold front. (These will be discussed later.) The bases of warm frontal storms are on the slope of the front, and the effect of the storm is felt primarily as heavy rain beneath that frontal slope. IFR flying conditions might not be too bad at lower altitudes. Higher, though, and above the slope of the front, the warm front storm might bang you around terribly. And where the frontal slope is close to the ground, as is true near the surface position of the front, the effects of the storm can be felt at low altitude. Best just consider all thunderstorms as bad.

Warm frontal slopes vary from 1:50 to 1:200 with the average at about 1:100. If you were 100 miles ahead of the surface position of the front, the slope would be just over 5,000 feet above the ground, so you can see that staying below the slope to minimize the exposure to embedded warm frontal storms means flying rather low. We aren't going to look at a drawing of a warm front cross-section, as the typical drawing of these always accentuates the slope to the point that it renders the illustration meaningless. Just visualize it as rising one inch in 8.3 feet to get an idea of how very shallow a warm frontal slope really is.

STABILITY

The stability of the warm sector air is a key to the likelihood of storm development in the warm frontal area. If the temperature drops gradually with altitude behind the surface position of the front, the air is relatively stable and thunderstorm development isn't likely ahead of the front. If the air tends to be unstable, though, watch out when it starts ascending the frontal slope. You might say that it is looking for trouble, and finds it ahead of the front.

Temperatures aloft are measured, and our most common exposure to them comes in the form of a forecast, included with the wind aloft forecasts (except for the 3,000 foot level). These should always be considered and compared with actual temperatures as you fly. If the temperature aloft is warmer than forecast, the air is more stable than expected; if it is colder, it is less stable. If the forecaster predicts the temperature will drop more than 2 degrees C per 1,000 feet, watch out. That is a sign of expected instability.

The distance from the low pressure center is also a factor in judging the possible severity of a warm front. If the low is many, many miles to the west of the point at which you'll penetrate the front, then you'll be in an area where the ascent of warm air is not so directly aided, abetted, and aggravated by the inward and upward flow associated with the low center. The meanest warm frontal penetration is closer to the low, just east of it. This is where you'll find the worst of all worlds.

WARM ICE

Though the terminology may make it seem illogical, the warm frontal zone is often pretty icy in the wintertime. Warm can mean warm in relation to something that is very cold, instead of warm in the literal sense of the word.

Ice can prevail north of a warm front because that's where the necessary ingredients come together. There's moisture coming up from the south; that moisture is being lifted and cooled as well as being transported to an area where it's just naturally colder. If freezing rain is reported or encountered, it is a good indication that there is warmer air aloft, above the air that is below freezing. This warm air above offers a haven from ice, but in some situations the above-freezing levels can be very shallow. For example, the temperature at the surface might be −4 C, and it might rise to +1 C at, say, 6,000 feet, which would be the slope of the warm front. Then it would drop off above that level at a rate of 2 or more degrees C per 1,000 feet.

The only time there is real salvation in warmer temperatures aloft is when they are comfortably above freezing. Also, we must consider that as we fly toward or away from a warm front at a constant altitude, we'll be continually changing our relationship with the slope of the front. In other words, what works over Springfield might not work over Tulsa.

Despite the importance of the warm front to the general aviation pilot, it is often given scant attention in weather briefings. The emphasis is usually on cold fronts, with the warm one often not mentioned as such. Instead it is considered just a big area of bad weather. But it is important for us to know the surface position of any warm front, and any time there is a low pressure center to the west we should seek that information. By knowing the surface position, we can get some idea of the approximate point at which we will penetrate the slope of the front at the chosen flight altitude, a bit of information that has a direct bearing on the effects of embedded thunderstorms. For example, if you were flying south at 5,000 feet directly toward the warm front, the maximum exposure to actual thunderstorm cell penetration would come in the 100 miles north of the surface position of the front.

Warm fronts are not dramatic. They lack the clear-cut and decisive candor of a brisk cold front, and this is one reason why a warm front isn't always depicted to the east of a low. The factors that define a front (a change in temperature, wind, pressure, and dewpoint) might not all change and justify a line on the map called a warm front, yet the weather that we encounter east of a low might have the look and feel of one. Even when there's a warm front on the map, we might have trouble pinpointing a surface position as we fly through it. The inclement weather might not actually begin for some miles ahead of the front. Regardless of any lack of drama, it still pays to be curious and view the area east of a low as one where there will be conditions similar to a warm front, even if there isn't one on the map. Lack of curiosity can lead to some unexpectedly wet and bumpy rides.

UPSLOPE

To the northeast or north of a low, the easterly flow can combine with terrain features to create very low clouds and fog. This is called an upslope condition and can occur anywhere the circulation is moving warm, moist air over increasingly higher terrain. The increase in terrain elevation provides enough lifting to cause condensation and cloud formation, and where the area affected is a large one of generally increasing elevation, the upslope condition can affect a very wide area. A really big hunk of the Great Plains can go to zero-zero in an upslope condition, and can stay that way for days—until systems move and the easterly circulation is modified.

COLD FRONT

Let us move now to the cold front. Here, cold air is moving in under warmer air. The slope of a cold front is steeper, but it still isn't all that steep. The average is 1:80; if you drew it, the cold air would be an inch deep 6.6 feet from the surface position of the front. It is important that the surface position of the cold front is at the leading edge of the slope, instead of at the trailing edge of the slope as with a warm front. The cold front is pushing and shoving under the warm air ahead of it, rather than sliding warm air above cold air as with the warm front. The result is a narrower frontal zone, with more action concentrated in the frontal area.

As with the warm front, the stability of the air in the warm sector is related to the severity of weather that might be expected in or ahead of a cold frontal zone. The less stable the air, the more likely the fireworks.

There are a lot of things to consider in trying to make a judgment on the severity of cold frontal weather. The width of the frontal zone can be one key. If it is rather diffuse, and perhaps a hundred or so miles through, things in there might not be so severe. If the frontal zone is narrow, with a dramatic temperature and wind change in a short distance, it's probably chaotic in the frontal zone. If there is a strong southerly flow ahead of the front and a very strong north-westerly flow behind the front, there will be a lot of convergence at the surface. All that air colliding is bound to create a lot of lifting. Thunderstorms are a strong possibility, along with generous amounts of turbulence. Even if storms do not develop, the turbulence caused by the convergence can be significant.

Most cold fronts will have some turbulence for us at lower levels. Topping them is relatively rare in light airplanes, although it is occasionally possible. It's best to expect bumps in any front; then a lack of them will be a pleasant surprise.

When really severe weather develops in connection with a cold front, it is generally found well ahead of the surface position of the front. This shatters the theory of the cold air pushing ahead and causing lifting, and thus storms, because those storms simply form far ahead of the push and shove at the surface. This is a direct effect of upper air patterns, discussed in the next chapter.

We can watch the surface wind behind a cold front and tell how the whole system is going. When the wind is strong and perpendicular to the front, frontal movement across the ground is continuing. When the wind tends to become lighter or shifts to the north or the northeast, to a direction more nearly parallel to the front, then the front is slowing or stopping. The basics of circulation tell us that in this case the low is weakening, and the primary influence is a high pressure center to the north or the northwest, instead of the low.

The speed with which fronts move is related both to the general easterly or northeasterly track of the low pressure system and to the strength of the circulation around the low pressure system. A typical cold front might average 25 knots when moving over flat country, and might slow down over rough terrain. Some go faster, some go slower, and the speed can always change.

When a low is gaining strength, circulation can be very strong and movement of the low can be quite slow. In this

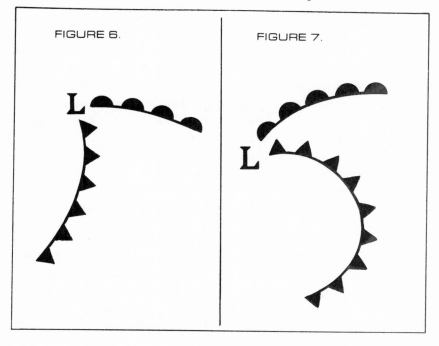

FIGURE 6.   FIGURE 7.

situation, the fronts can move ahead of the classic position that is normally pictured. Instead of appearing as in Figure 6, they would appear as in Figure 7. This can continue for only so long, however. The low will eventually get moving, or the usually faster-moving cold front will catch up with the warm front, move under it, and form an occlusion with the leading edge of the warm front aloft instead of at the surface. The whole system reaches a zenith as the cold front starts to overtake the warm front; from there it is all downhill, as the low surrounds itself with air of about the same temperature and, lacking the low level feed of warm moist air, starts to weaken.

When the situation in Figure 7 does develop, the front a great distance away from the low might not be so violent, but you can find the potential for violent weather close to that deepening low pressure center. Visualize the circulation around it, too, and you'll see that the surface wind behind the cold front will be from the southwest, instead of the west or northwest as we normally anticipate. The wind will likely be very strong close to the low, too.

PRESSURE

A front itself is an area of low pressure. Note in Figure 8 that the isobars (lines of equal pressure) show that the pressure will drop as the front approaches, reach a low reading as the front passes, and then start increasing after frontal passage. The pressure jump behind a cold front might be more pronounced than that behind a warm front, but there is still a perceptible pressure increase with warm frontal passage.

Nothing lasts forever and that is certainly true of fronts and low pressure systems. A cold front can move only so far south before the low moves far away and weakens, leaving it without push. Also the cold air is warmed by the surface over which it flows, and becomes similar to the air on the other side. Some fronts dissipate, others become stationary. Basically, though, there is always cold air to the north, warm air to the south (all in the northern hemisphere), and the line between the two undulates as the world's temperature mechanism works to maintain some sort of balance. When the south-moving cold

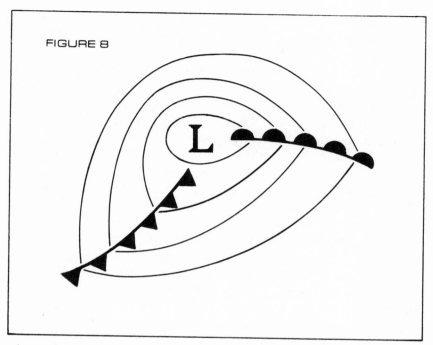

FIGURE 8

air pushes as far as it can go and stops, that cold front can be considered at an end—but its relationship with warm moist air becomes the breeding ground for the next weather system.

As we discussed before, the phenomenon of a low pressure storm system developing on a stationary front can be at its classic best in the Gulf states, probably most pronounced in January and February. In December, cold fronts often whip through this area and carry cool air into Florida. But as the days start getting longer, there is more resistance to the southward movement of cold fronts. A front might grind to a halt off the east coast, and extend down across south Georgia and then southwestward into the Gulf of Mexico. When this happens, there is cool northeasterly flow on one side, warm southwesterly flow on the other side, just waiting to cause trouble.

Some weather folk in the south have a nice term for what happens next. They call it a "return cold front." A cold front passes, goes south, and then comes back. The return visitor is actually the bad weather north of a warm front, but the "return cold front" terminology is rather nice.

What makes it return? It is a simple case of low pressure generation along a stationary front, as in Figure 9. There is northeast wind on the cold side and southwest wind on the warm side. A wave develops, as discussed in the preceding chapter, and curls all the way over, forming a low center with complete circulation. In this illustration of a typical Gulf low, the system moves slowly while developing and gaining strength and then moves quite rapidly up the east coast. The front is cold to the southwest of the low. The warm sector is at sea. The primary effect might be felt as a northeaster, the famous storm that rakes the east coast with rain and snow-storms. If the storm forms far to the west and moves south of the Gulf states, it can bring storms to that area.

There are four spawning grounds in particular that frequently develop good strong low pressure storm systems: the Gulf, the area offshore Cape Hatteras, Colorado, and south-western Canada. Lows form in these areas and, given complete development, they are long gone from the North American continent before they begin weakening. We get the whole nine yards from them.

KEYS TO DEVELOPMENT

Weather forecasters are able to do only an approximate job of forecasting the development, track, and intensity of these low pressure systems. As we'll see in the next chapter, the upper air patterns trigger and steer these systems, and it is difficult to tell what's going on up there from minute to minute, and to detect shifts that can have a dramatic effect on low development. Also, the Gulf and Hatteras lows are difficult because there are no surface weather reports in the areas where they form. This means that pilots must use a knowledge of the basics of circulation to supplement the forecasts.

If overrunning warm air is evident (in the form of high clouds moving in from the southwest) and you are to the east of a location where low development is suspected, beware. The sailors of old used a mackerel sky as a sign of impending meteorological doom. This was their description of high altitude cirrocumulus or altocumulus clouds forming in basically

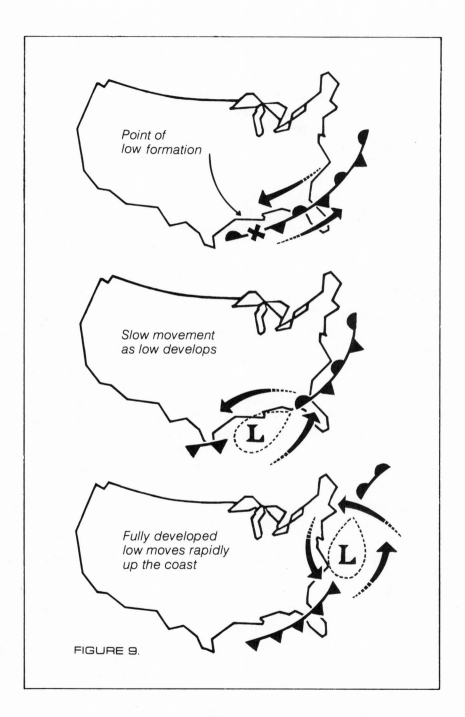

Point of
low formation

Slow movement
as low develops

Fully developed
low moves rapidly
up the coast

FIGURE 9.

clear weather on the warm frontal slope a great distance ahead of the front—a sure sign that the air ascending the slope is unstable.

Another key might be an increase in precipitation in generally grungy weather to the north of a stationary front. Any increase in surface or low level winds from a southerly or southeasterly direction is also a clue.

After a low has formed, the surface winds offer good clues about its track in relation to a given location. Southwest winds mean that the low is likely to pass north of your position; south to southeasterly winds mean that it might pass close to your position. Easterly surface winds suggest a passage south of your position.

Thinking in terms of surface wind, the more easterly the wind, the more likely you are to get the worst of a low as it moves. The weather then might well be at its wettest and last the longest. In the eastern half of the U. S., the southerly and easterly flows simply mean more moisture. The more westerly the wind, the more likely you are to have generally acceptable flying conditions, except in thunderstorms related to a squall line or to the cold frontal zone.

Also, remember the basic that a low doesn't move much when gaining strength. A really good one has to spin up before it takes off across the countryside. It might take the low a day or so to develop and then start moving. Again, it is nurtured and steered by upper level wind patterns, which we'll examine in the next chapter.

ANYWHERE

Generation of lows can certainly occur at places other than the four most common areas, and we sometimes see the generation of more than one along a stationary front. As things are developing, the map might show several lows in a general area. However, a very strong low will use up moisture and energy, and any lows on a map with another really strong one within 500 miles aren't likely to grow up to be big and mean and have fully developed frontal systems. They often disappear when the strong one gets its act together and starts moving.

Sometimes when studying weather we become mesmerized by fronts. And, indeed, fronts are the most convenient mechanisms for discussion in a textbook, or in a weather briefing. We need to think of lows, however, because if operations are conducted in that big segment that is north of the warm front and north of the cold front, there might be plenty of bad weather unrelated to any nearby surface front. Understanding cold and warm fronts doesn't do much good up there; the predominant effect is from the low itself, as noted in Figure 10. There, the movement and location of the low, not the movement and location of the fronts, is the key.

The large-scale ascent of air is predominant ahead of a moving low. True, there might be an identifiable warm front somewhere in the circulation ahead of or east of the low, and the air might be moving up over this, thus relating it to the front. But up around the north side of a low, far from any warm front, you can find some of the worst weather going. Blizzards along the east coast are usually found in this area. The strong east winds bring in moisture, the low provides lifting, and cold air to the north mixes in to make it all nice and frosty. In the springtime, a generous supply of warm and moist air at the surface can combine with a high level of instability in the path of and east of the low to make this a spawning ground for severe thunderstorms.

It is true that the farther around the top of the low you go, the less the chance of encountering really major weather problems. As the moist air swirls inward and upward, it loses its moisture and there's less and less stuff with which to make clouds and precipitation. But that certainly doesn't mean that it will go VFR as soon as a low passes, or even within a reasonable period of time after. In many cases, the moisture supply is strong enough to generate clouds all the way around on the west side of the low. Perhaps the tops will be reasonably low, and there won't be a lot of precipitation, but enough clouds can persist to harass at least a VFR pilot.

The relationship of the surface low to any low aloft is important in positioning weather around a low, too. We'll explore that in the next chapter.

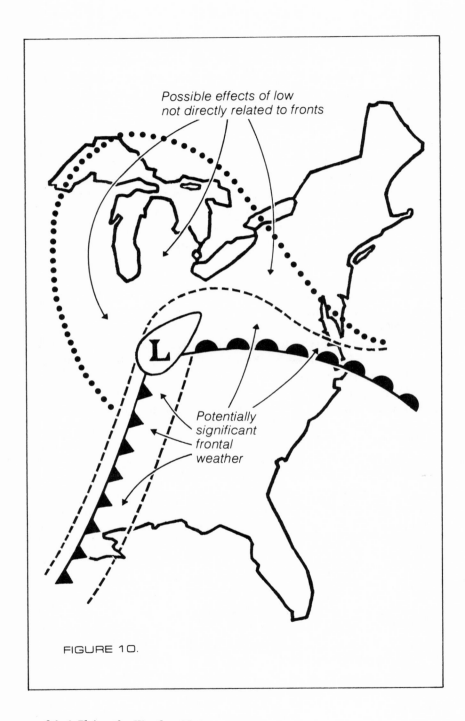

Possible effects of low
not directly related to fronts

L

Potentially
significant
frontal
weather

FIGURE 10.

TROUGHS

We sometimes hear troughs mentioned when listening to weather broadcasts or getting a briefing. These tend to come off as rather mysterious phenomena. Indeed, one wise old pilot once whispered in my ear that any discussion of a trough is a tipoff that the weatherman doesn't know what's going on, and that it is time to watch out for unforecast trouble.

A surface trough is identified by a bulge in the isobars away from the low where there is not a surface front. The significance of a surface trough can vary from nothing to considerable, and it might be safest to consider a trough as a front until it proves itself otherwise. The pressure is low in the area, proof of some convergence at the surface and divergence aloft. Given enough of both, some very interesting things can happen. Otherwise unexplained collections of thunderstorms are found in troughs, for example. If you don't want to think of a trough as a front, think of it as an extension of the low pressure system, with bad weather to the east and improving weather to the west. Also, when there are two lows on a map, consider that there is usually a trough between them. Connect the two lows with a line and, even though no specific map feature suggests action, don't be surprised if there is some weather along that line.

DOUBLE COLD FRONTS

Two cold fronts occasionally trail from a low pressure system. In this case, the circulation is usually very strong, with the air behind quite cold and the air ahead quite warm. The first front doesn't quite get the temperature down to the potential of the total circulation, so a second identifiable front forms behind it to complete the icy blast.

MOUNTAINS

Terrain can have a very definite effect on the movement of low pressure systems and the composition of fronts. There's drag in rough terrain and, just as aerodynamic drag slows an airplane, rough terrain slows the progress of a low or a front. It can even stop a front that isn't moving very rapidly. By the same token, once a front is past mountains, it can start moving more rapidly over the surface. Mountains have a more pronounced effect on cold fronts (it is easier to pull something over a rough surface than to push it) and this can result in poor weather lingering longer over rough terrain.

Mountains can also create conditions that seem like fronts to us even when there is no frontal activity. When there is a strong northwesterly flow over the Appalachians behind a front that is well off the Atlantic coast, a very impressive array of clouds and snow showers, with tops to 20,000 feet, can develop along the mountains. For this situation to develop, the ground must be warm relative to the air passing over it, contributing to a basic instability. The deflection of the wind by the mountains contributes to lifting, and the combination of moisture from over the Great Lakes and moisture being brought over the top of the low pressure center provides the stuff of which snow is made. It might well be clear to the west and clear to the east, with a seeming wall along the mountains, 100 or more miles through. It's no front, but flying through it is somewhat similar to flying through a front.

The Great Lakes can do crazy things to the weather, too. After a cold front passes, the lakes are warmer than the air, creating enough instability to trigger showers that form over the lakes and move inland for a number of miles before dissipating.

REVIEW

In starting a review of fronts, let's get rid of what might be classed as a myth. We often tend to look at the surface weather chart and consider fronts as walls that go straight up. We look at the surface position of the front and think of that as its location. This is true only if you are traveling on the surface. Flying, we'll find the front in quite a different position. The slope of any front is shallow and, even at the relatively low altitudes used by most general aviation IFR airplanes, the front aloft can be 200 miles or more from the surface position of the front. So let's think in terms of frontal zones instead of clearly defined fronts. Instead of lines on maps, think of bands of weather. The warm frontal zone, the area where the weather is directly affected by the warm front, is a blob that's usually to the east and the northeast of the low. Decreasing clouds come around the top of the low and curl behind the cold front, which can be either a broad and not too bad area, or an intense and rather narrow area generally running off to the south or southwest of the low. Think too about the weather in relation to the low, without specific regard for the fronts. If the low passes to the south of your position, you get good mixing of the moisture from the south and the cold air from the north. Icing can be a big problem. If a low is southwest of your position moving to the northeast, you are in its path. If the low is to the west, you might be affected by the classic warm front. If the low is to the northwest of your position, you are probably already in the warm sector and will miss the effects of the warm front. Instead, you have only the cold frontal zone to contend with. If the low is east of your position, things should begin improving sooner rather than later.

Those are things to contemplate before takeoff. When aloft, we don't think of weather relative to a stationary point. The airplane is dynamic and it moves faster than the weather, so we study the relationship between two things that are moving. We have a basic advantage here, too, in being able to fly the airplane in a manner that minimizes the effect of the weather. On the ground, we sit helpless as a thunderstorm rumbles overhead; in the airplane, we hopefully go around the fury of the storm. That's weather on a small scale. On a larger scale, it is often possible to adjust a flight path to minimize the effects of weather. We might go southwest for a while before turning to the west, to get through a cold front at a better spot a bit farther away from the low. Or perhaps a detour to the south will enable us to conduct the flight in the warm sector.

Knowing what happens as the low passes a point lets us form an opinion of weather to come, so it's important to check weather continually while en route. Study the weather sequence reports for locations ahead of and behind the low, to get an idea of how things are shaping up. An example of this is as aviation-related as boat-related, so it's worthy of consideration: Lake sailors in the middle of the country use the transcribed weather reports to judge how the wind will develop on the lakes in the course of a day. If wind velocities at reporting stations to the west and southwest are within limits for the sailor, then the wind on the lake is not likely to exceed limits in the near future. That can be applied to aviation equally, for both wind and weather. See what it's like in the direction from which the system is moving.

However it is sliced, we must know our position in relation to the low pressure area (or areas) and fronts as we fly. If a pilot doesn't know position in relation to weather, he or she is as lost from a weather standpoint as a pilot is lost navigationally when ignorant of position in relation to points on the ground.

Upper Air Patterns

Most of the flying weather information we get is related to surface conditions. The map in the flight service station is the surface chart. The weather sequence reports we get are surface observations and the terminal forecasts are educated guesses on expected surface conditions. As superficial as it might seem after we examine the effect of upper air patterns on weather, this system has withstood the test of time rather well. Still, we must recognize that what goes on at higher altitudes has quite an effect on flying weather at the lower altitudes and on surface weather. Forecasters themselves make extensive use of upper level information. And while every pilot sure can't be an expert meteorologist, knowing a few basics can help us better understand what is going on, as well as avoid surprises. Weather is very definitely three-dimensional, and we must know a little about the effects of upper level pressure and wind patterns to find logic in forecasting and to examine difficult situations. The pilot who doesn't have the basic knowledge to openly examine and question every forecast is ill-equipped.

RULE-OF-THUMB

The 500 millibar chart is a widely used upper level chart. The 500 mb chart shows the height of the 500 mb pressure level, which is approximately 18,000 feet above the surface and which varies with pressure. Lines tracing the height of the 500 mb surface are drawn on the chart and are used like isobars— lines of equal pressure—on a surface chart. The wind flow at altitude is parallel to the lines of equal height. The charts include relative humidity, wind direction and speed, temperature, temperature and dewpoint spread, and the change in height of the pressure level in the past 12 hours. Charts are drawn for other levels, based on actual upper air observations, but the 500 mb chart is the most significant, especially for the pilot interested in basics.

The patterns at the 500 mb level can reflect closed circulations around lows and highs at that level, meaning that the circulation goes all the way around the pressure area, as in Figure 11a. Or they can show an undulating west-to-east flow, as in Figure 11b; or they might be a combination of the two. The southward projections are troughs with colder air often circulated down around the tip of the trough where it then heads back north. Figure 11c shows lines of equal temperature, as related to the pressure patterns in Figure 11b. Air at 18,000 feet is not directly affected by the surface temperature, so it is not modified in the southward trip and remains cold.

The movement of this cold air over warm air below produces instability, so the 500 mb chart is a real key. Just remember that the situation is a bit reversed aloft—low pressure relates to cold, high pressure to warm. Cold air aloft over warm air means instability. Warm air aloft over cold air means stability. All temperatures are, of course, relative.

Basic: When there is a trough or a low over or to the west of you on the 500 mb chart, it means that there could be relatively cold air aloft at your position. If the air at lower levels is warm, that means instability—cold air over warm air will do it every time. That is the basis of a rule-of-thumb once furnished by a very wise meteorologist: If there is a low or trough on the 500 mb chart to the west of your position, be prepared for things to get worse. Also, at the tip of a trough aloft you

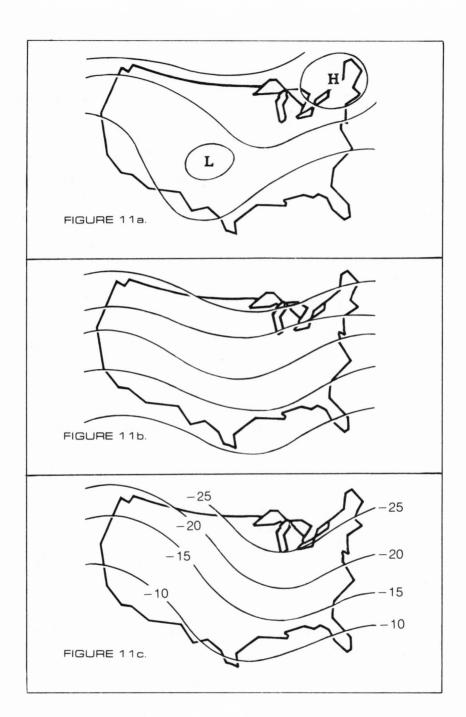

FIGURE 11a.

FIGURE 11b.

FIGURE 11c.

−25

−20

−15

−10

−25

−20

−15

−10

might find the maximum southward circulation of cold air. Again, cold above warm makes instability, and the tip of the trough becomes a likely spot for low development.

Let's examine why a low pressure area at 18,000 feet suggests trouble to the east.

In Figure 12, we have a couple of lows on the 500 mb chart. The lines of equal height are close together and the winds aloft stronger to the southeast of the lows. This particular circulation was moving cold air down around the south side of the low over Arizona and then turning that cold air back to the northeast. The surface weather chart positioned a primary low up in Wisconsin, with an occlusion forming northeast of the low and a cold front back down through central Oklahoma and Texas to a low in northern Mexico. Surface temperatures were about +22 C ahead of the cold front's surface position, with an abundance of moisture pulled from over the Gulf by the southernmost low. There was an outbreak of strong storms, with hail and tornados, in Texas.

The weather-wise person who told me the theory about lows to the west causing trouble made the explanation as simple as possible and it can be seen readily on this chart. Transporting that cold air from up north around the south side of that low aloft, over warmer air at the surface, meant instability, the likelihood of more vertical development and more storms.

Consider too that the ascending air curling up around that southernmost surface low would just keep right on curling into the low aloft, as could the ascending air in the northern low. That's what you call good upper level support for storm system development. The rising air in a surface low has a place to go. We often hear real meteorologists use those words "upper level support." It is an important part of the equation.

The upper wind patterns tend to undulate, and the undulations move across the country. Note the 500 mb charts in Figure 13. On Monday, the lines are almost straight east-west with a little southerly pooch, a developing trough, in the western U. S. On Tuesday that trough is more pronounced and has moved to the east. On Wednesday, a low center is noted over the north-central U. S. and that trough is even more pronounced and a bit farther east. The eastward motion has progressed some more on Thursday.

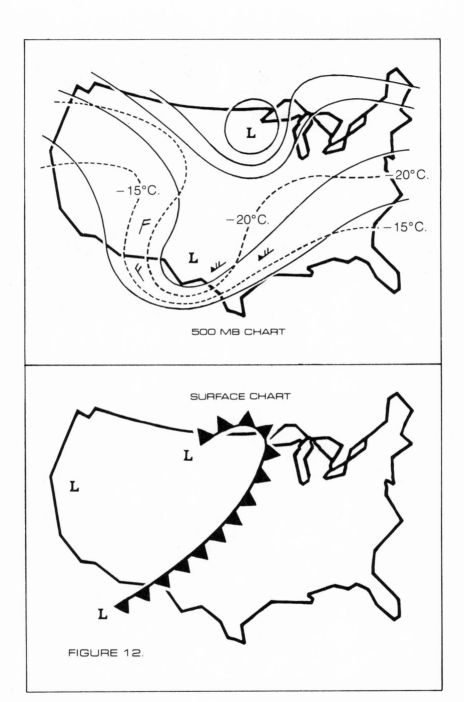

500 MB CHART

SURFACE CHART

FIGURE 12.

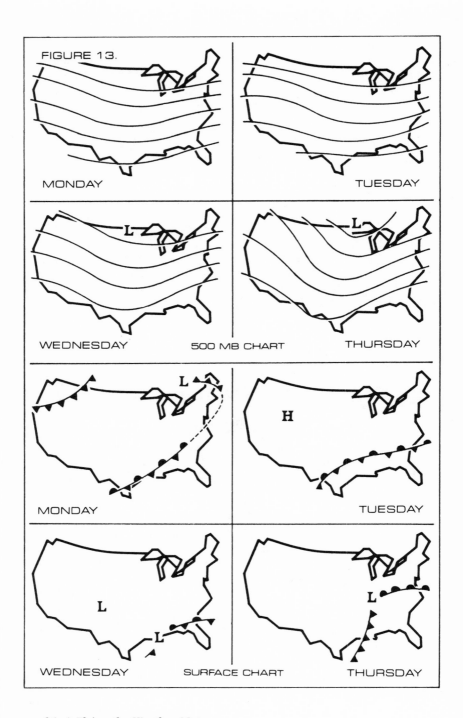

FIGURE 13.

MONDAY

TUESDAY

WEDNESDAY   500 MB CHART   THURSDAY

MONDAY

TUESDAY

WEDNESDAY   SURFACE CHART   THURSDAY

Next look at the surface charts. On Monday there is a weak low in the northeast. The cold front has become diffuse and is drawn as a stationary front to the southwest. This low has lost its upper level support and is losing its identity. The circulation around it is slowing. The winds above it are straight east-west. The front in the west is weak. On Tuesday a big high holds sway in the west and Monday's dying front is stationary across the southeast. On Wednesday two lows have formed. Note that they formed in relation to the southern tip of the trough on the 500 mb chart. The stationary front at the surface provided a ripe spot for the development of a low, but it had to wait until Wednesday for the upper level circulation to provide some support. On Thursday, one of the lows has developed into a strong one and has moved northward with an active frontal pattern. Note how the low at the surface tracked the southwesterly flow east of the trough on the 500 mb chart.

The formation of the low at the tip of the upper level trough, and the track of the low to the northeast or north-northeast, below the circulation aloft, is a standard pattern.

While the effect of patterns aloft on surface patterns is relatively easy to predict, the behavior of the patterns aloft is quite difficult to predict. The depth of the trough, and the distance from ridge to ridge (a ridge being the opposite of a trough, or a northerly pooch in the lines of equal height on the 500 mb chart), have a bearing on how rapidly the waves move across the country, but it is still difficult to forecast the precise development and track of upper level low pressure areas and troughs. That is work for real meteorologists. For our basics, it is enough to note the development and position of the lows and the troughs on the 500 mb chart.

JET STREAM

The core (the area of the strongest winds) of the famous jet stream is usually higher than 18,000 feet, but it is still a dominant feature of the circulation on the 500 mb chart. It also plays an important role in frontal activity and the development of severe weather. The jet stream moves parallel to the lines of equal height aloft, and it moves faster than the troughs. For example, using Figure 12 (page 35), on Monday a jet would be barely on the map, at the upper left corner. By Tuesday it would be working toward the southeast. On Wednesday it would be at the southern tip of the trough and by Thursday it would be northeastbound, ahead of the trough. The jet stream is often at a higher velocity when east of the trough, northeastbound, as on Thursday. In this particular situation, 85 knots was the strongest wind at the 500 mb level until Thursday, when it reached 105 knots over eastern Tennessee.

There is a circulation within the jet stream that is offered as a direct cause of severe weather in the warm sector ahead of cold fronts. This circulation results in descending air in the right front quadrant of the elongated, kidney-shaped jet stream, as in Figure 14. The compression of this descending air results in heating, which accounts for unusually warm air ahead of the surface cold front which, in this illustration, would probably be below the center of the jet stream. The descending air means it is probably clear in this area, just southeast of the jet stream.

This warmer air pushes out to the southeast, ahead of the cold front and into the warm sector. The push results in convergence 100 miles or so ahead of the cold front, and basic instability associated with the warm sector air at the surface and the cold air aloft to the southeast of the jet stream can provide the ingredients for rapid and dramatic thunderstorm development. This also explains why weather often clears after passage of a squall line and remains clear and relatively warm until the cold front comes along.

Tracking the jet stream is a very important part of forecasting severe weather; for us it is enough to know that when it's northeastbound within a couple of hundred miles west of our position, the thunderstorms are likely to be both in lines and quite severe.

The erratic movement of the jet stream is good reason for not judging weather by what happened the day before. Activity in a frontal zone might be garden-variety one day because of only moderate upper level support and no jet stream effect. The next day the jet might have moved down around the trough and be moving northeastward, generating the very real possibility of extremely severe weather.

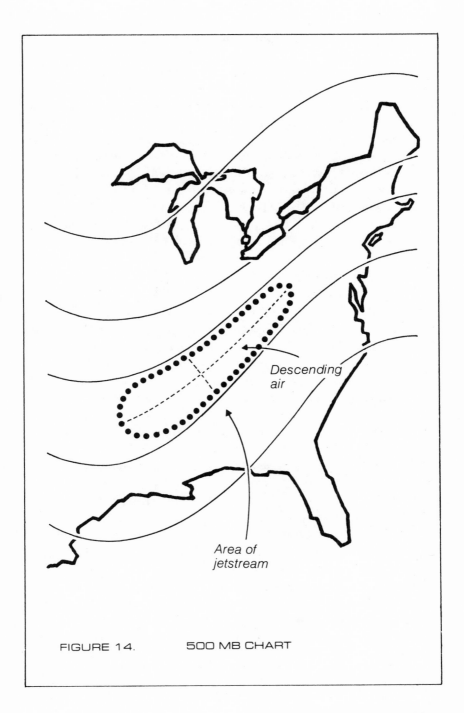

Descending
air

Area of
jetstream

FIGURE 14.   500 MB CHART

If northwest of the cloud shield of a rather clearly defined but very slow-moving cold front, you see delicate high clouds moving speedily to the east or to the northeast, that is likely the jet stream moving around and heading northeast, ready to make mischief. As it comes around, it can actually draw the front back to the northwest a bit if everything is just right. But the most notable effect will probably come a day later, after the jet stream gets headed to the northeast and does its thing to add additional instability to the warm air east of the cold front. This phenomenon was very noticeable in the weather system that spawned the extremely strong storm that downed a Southern Airways DC-9 in 1977, as well as in the system that devastated Xenia, Ohio, and a lot of other territory in 1974.

CLOSED LOW

When the circulation aloft becomes complete around a low aloft it is called a closed low, and if the circulation to the north of the closed low flattens out into an east-west direction then it is a cut-off closed low aloft—more likely to slowly dissipate than to move. With a closed low, upper level support for a surface low is strong in the area of that upper low, as well as at the southern end of the trough aloft. If the upper low is cut off, then there is likely to be an extended period of bad weather below and to the east.

Closed lows aloft can also cause unusual movement in surface systems. A classic example of this came with Hurricane Agnes, in 1972. A relatively rare June storm, Agnes moved up the east coast to a position off Cape May, New Jersey. From there it would normally be expected to go on north or northeast, hosing things down along the way. But a deep closed low aloft had developed over Pennsylvania, and Agnes, seeking some place to go live a little longer, moved westward to a position under the closed low. The hurricane winds of the storm subsided when it moved over land, but the low was still strong and the upper level support was almost absolute. Record flooding in Pennsylvania was the result.

When the lines of equal height on the 500 mb chart are east-west, without troughs or ridges, there is no guarantee of good weather. But this is a sign that there won't be a lot of upper level support for any system that tries to develop. Surface lows and fronts are not likely to extend their influence up to a very high altitude.

Do remember that when considering the upper air patterns, we are looking primarily at temperature relationships, and stability, and that the seasons have a lot to do with our interpretation of upper air charts. In the winter, a flat, east-to-west flow on the 500 mb chart means that strong low development is unlikely. But in the summer, the appearance of a flat east-west flow at 500 mb does not preclude localized outbreaks of thunderstorms, often severe thunderstorms. In summer, surface temperatures are relatively a lot warmer, and instability can be greater. Slightly cooler pockets of air can also move eastward in the upper level flow and, when one passes over an area, it does its bit for instability. The comparison between temperatures at the surface and aloft is what we want to watch, along with any clue to lifting that might trigger the development of convective activity.

BASIC FLOW

It is often difficult to get much information on the 500 mb chart from a flight service station. Many of them don't even keep the charts, simply because a lot of FSS people neither understand nor appreciate their importance. When they do have them, it is often difficult to get a good explanation of the chart on the telephone. In either case, it is still possible to get some idea of events aloft by checking the wind and temperature forecast on up to the 18,000 foot level and comparing this basic flow with the surface chart.

If there is a strong southwesterly flow forecast aloft over or west of your position, that means you are somewhere east of the trough aloft. If that is the case, find out if there is a strong surface low pressure system shown. If so, where is it? If the low is to the west or northwest, moving northeast, the pattern should be reasonably well covered in the forecast. The low has probably formed and is moving up the east side of the trough as they will do. Even if there is no clearly defined surface low when you see that strong southwesterly flow at 18,000 feet, one might yet form and do crazy things to the weather. It is difficult to predict the exact time of formation, so forecasts should be examined with suspicion. If the situation is one where a low has just formed and is southwest of your position, again with that strong southwesterly flow aloft, then chances are that the low will take a bead on your location.

Secondary lows can certainly form, and as a trough moves across the country at the upper levels it can spawn and steer more than one low pressure area. Indeed, when the classic northeaster storm forms off Cape Hatteras, there is often a strong low pressure center over in the Ohio valley. But when the Hatteras low forms and develops, the Ohio low weakens. It might be said that most upper level troughs spawn and steer only one strong low pressure area at a time during their crossing of the U. S.

Often, we can be more than a little perplexed by surface low positions from one check of the weather to the next. For example, the briefer might position the low in eastern Tennessee when you call in the early morning; later, when you check weather en route or at a gas stop, they might be talking of the low in southern Mississippi. The primary message here is that the system as it will dramatically affect flying weather is still in the formative state. Once the primary low forms and starts moving, the forecaster can do a pretty good job with it. Before then, the forecaster is looking more at possibilities than facts.

LONG TERM

When an upper level pattern does spawn a series of surface lows, it is usually because things get into a rut, resulting in extended periods of uniformly bad (or good) weather. In the record-breaking cold winter of 1976–77, for example, the eastern U. S. was persistently on the west side of the trough, which brought on very cold and dry air as a primary fare. The blizzards of 1978 came when lows aloft kept forming in the west, developing and steering surface lows along the snowstorm tracks.

A blocking high pressure area or ridge causes stagnation of the patterns aloft. If there is a big stationary high to the east of your position, that means the wind aloft over your position will remain southwesterly. Low centers will develop and move northeastward. If the blocking high is to the west, you'll be in the northwesterly flow, cold and dry.

LOW LEVEL CIRCULATION

The circulation at levels above the surface but below 18,000 feet can tell us a lot about the weather. Even when the low level circulation is more likely a result of other things, not a cause of things such as the circulation up higher, it is still important in putting together a picture of the flying conditions you can expect.

A good example can be found when there is a light northeasterly surface flow over a wide area, with southerly or southwesterly flow from about 1,000 feet on up. The surface flow suggests the presence of high pressure to the north; the southwesterly flow is warm air overrunning the cold. There might not be a warm front defined on the map to the south, or a low pressure center to the west, as the circulation would suggest. If you stick around long enough, however, something like that will probably develop.

Relatively strong surface winds and diminishing winds with altitude are an indication of a clearly defined and strong surface low somewhere around, but without much upper level support. The same situation would exist when the surface winds are southerly or southwesterly, with the winds aloft shifting to westerly or northwesterly by the 9,000 foot level.

When there's a snowstorm to be dealt with, icing in the clouds is more likely to be associated with the southerly and southwesterly winds. Once the winds veer around to a more northerly direction, the temperatures aloft are probably colder, with less chance for airframe icing. That's just one more example of how we can stay ahead of things by being aware of circulation and temperature.

WHO IS ON FIRST?

As with any cause-effect relationship, I suppose that the interface between upper air patterns and surface weather could be reversed and a case made for the surface patterns steering the upper winds. The variations in upper air temperature could be related to the lifting around surface lows. But the evidence is strongly in favor of the upper influencing the lower, so we should accept it that way. Too many of the alleged mysteries of weather disappear when we apply the third dimension and consider the influence of upper air patterns on surface weather.

Our airplane is a super weather sensor, too. A person on the ground can only apply his or her observation of surface conditions to weather deliberations, where we can use our airplanes to measure winds aloft, temperatures aloft, and the nature of clouds—bumpy cumulus or smooth stratus. We really *experience* the weather, as our airplane provides a grandstand seat for the greatest show above the earth, the elements on parade.

Chaos in the Sky: Thunderstorms, Wind Shear, Turbulence

After we charge our brain with some basics of weather in general and flying weather in particular, the next step is to consider some detail of such small-scale but important phenomena as thunderstorms and wind shear, and to consider precipitation's relationship to turbulence.

BEWARE THOR

The scripture says, thou shalt not fly into a thunderstorm, and it does so for very good reason. The action in a storm can be more than an airplane can take, or at least it can be too chaotic for us to remain in control. Indeed, even big airplanes have thunderstorm problems. But all thunderstorms are not alike. There are big ones and small ones. Some are visible for miles, some are embedded in other clouds. Some form in clusters, some in lines, and some go it alone. It is a broad subject without totally definitive answers. Only a couple of things are certain: In light airplanes they are bad, or worse; and there is always a good reason for storm development, or for lack of development, in what seems like a ripe area.

A thunderstorm (proper name, cumulonimbus) starts off as

a cumulus cloud. I shy from textbook methods and don't offer a picture. You know what one looks like. It's a head of cauliflower.

The feel of a cumulus cloud is as important as its appearance. It is bumpy inside a cumulus. When we are flying IFR in clouds, and it is bumpy, the clouds are cumulus. There is instability and vertical development.

As we plow through cumulus we can experience the nature of the clouds. If the bumps are sharp jabs that result in momentary airspeed excursions and make us work a little to keep the wings level, either the cumulus is just getting started, or we haven't gotten to the middle of it. If we find a good updraft, as shown by an increase in airspeed and a definite upward trend, then it's building at a pretty good clip and we have found the middle. In the building stage, all the action is up. It could build into a thunderstorm. We should be suspicious.

What starts the cumulus that grows into the cumulonimbus? The ingredients are well known: lifting action, instability, and a moisture supply are the necessities.

Let's first look at the simplest cumulus development, that of a hot summer afternoon. Some days there is moisture, as is evident by high humidity; and lifting, as is evident by the thermals; but no storms develop. We see only a good collection of puffy clouds. The missing ingredient in this case is instability.

The air is unstable when it cools more rapidly than 2 degrees C per 1,000 feet, and on a lot of hot summer days it just doesn't do that through very much of the atmosphere, at least not through enough of it to support the development of a cumulonimbus. Yet another day might seem just the same on the surface—hot and humid—and there might be an outbreak of storms even though there is no obvious mechanism, such as a front, to trigger them. What's happened is a change in the upper air pattern. A low or trough aloft, a cold pocket of air way up there, probably moved over the warm air and provided instability aloft. On days without storms, the cumulus tops built to 10,000 or 12,000 feet and stopped because the stability of the air aloft resisted further growth. The air just didn't cool rapidly enough with altitude for the cloud to stay warmer enough than the environment and keep building. But with cold air aloft, the warm moist air bubbling in the cloud gets fresh encouragement and keeps on building because air warmer than surrounding air tends to bubble right on up, to rise, not at all unlike a hot air balloon. The cold air aloft might not be on a large scale, as required to help support a strong low pressure system, but it might be large enough to support a good cluster of thunderstorms.

An individual thunderstorm cell does not last a long time. After the top builds through the freezing level, and once rain starts to fall, the storm moves through maturity and into a dissipating stage. But what we see as one big storm might really be a cluster of cells, in various states of generation and dissipation. A cluster can last for quite a while if conditions are good for development of anything other than isolated thundershowers, and the cells will continue to develop until a necessary ingredient disappears. In summer, this is usually lifting. When the sun sets, the thermals stop and the storms go away.

Moisture supply is also a key to storm development. It takes an enormous amount of water to construct a cumulonimbus, and if the moisture supply is a touch short, there might be enough for one, or for two, but not enough for a continuous development. The low level wind is the feed of the thunderstorm, so the strength and direction of flow plus the amount of moisture in the lower levels is a key to storm development.

A logical question: If there's lifting and instability but no moisture, why don't we get some cloudless bubbles of warm air rising to great heights, with all the attendant turbulence of a storm, but without the rain? Well, thermals do rise for a while without making clouds, but these don't go high and they don't become such a big thing because they are basically dry. Remember that the cooling of air is affected by moisture content. Dry air cools more rapidly than humid air, and thus it assumes the temperature of the surrounding air after it has risen for a relatively short while. On the other hand, moist air cools more slowly and as it rises it tends to stay warmer than surrounding air, maintaining its instability.

This is all easy to see on that hot summer day. A thermal starts bubbling up as the sun heats the surface unevenly and creates instability in the lower levels. The thermal is invisible until it reaches the condensation level, where the temperature equals the dewpoint, and cumulus clouds form. This is usually from 3,000 to 5,000 feet above the ground on a summer day. The little cumulus either builds and amounts to something or spends its life floating around as a puffy little fellow, decorating a pretty blue sky. It depends on, among other things, instability —the temperature drop with altitude—from the condensation level on up.

The temperature aloft is about the same, day or night. The sun's heating warms the surface during the day, creating low level instability to start the cumulus process. Without this, no small clouds form. That's why we have maximum clouds in the afternoon and minimum clouds at night.

The development of thunderstorms along mountain ridges is a variation on the hot day theme. A bit of wind over the ridges can cause the lifting action to trigger cumulus development, or heating in the valleys can cause a flow up the mountainside. (You can get an almost dry thunderstorm in the mountains if the lifting and instability are strong enough.) And even though it isn't as clear cut, the development of storms in connection with fronts or low pressure follows the same script. The big difference is in the front or low causing a large-scale ascent of air, as opposed to the hot summer afternoon storm's birth from small-scale thermal activity. Also, storms are often caused by a combination of heating plus lifting from some other source, such as a front. Either factor alone wouldn't do it, but together they might turn the trick in a spectacular manner.

FRONTAL STORMS

In examining warm fronts, we noted that instability starts on the frontal slope, and that thunderstorm development is from there on up, with the storms embedded in other clouds. Where this is true, the strength of storms may be mitigated to some extent. While the air in the developing cell is moist and thus cooling slowly with altitude, the same thing is true of at least some of the air around the cell. Thus the instability might not be as strong as where a cell is building through drier air that is cooling more rapidly with altitude. At some point, however, a warm frontal storm may build into colder and drier air; when one does, it can develop considerable fury.

The storms that form in the warm sector, south of the warm front and ahead of the cold front, have the best deal of all. The initial lifting related to the most severe storms is often caused by low level convergence well ahead of the front, triggered by effects of the jet stream. It is both warm and humid in this area of convergence, and the low level circulation is strong, reinforcing the moisture supply for storm formation. To top all that, a trough or low pressure system aloft to the west, common with strong cold fronts, especially in the springtime, can cause that advection of colder air southward around the tip of the trough where it starts to move back north, over the area where the lifting and moisture supply are getting things going. This creates everything in spades, and gives us some of the most awesome storms of all.

Garden-variety cold frontal storms can get a start from the lifting that occurs as the colder air pushes under warm sector air at the surface. These storms form in the frontal zone and can often form lines that are as impenetrable to light airplanes as the severe squall line that devlops well ahead of fronts.

NOCTURNAL THUNDERSTORMS

Nighttime thunderstorms that occur with no apparent lifting mechanism, such as a front or wind over irregular terrain, lack the clear-cut causes that we've associated with other storms, but if they are there, something causes them.

The air near the surface usually cools at night, but aloft things remain very much the same. If there is movement of warm air from the south in the lower levels, and if there is cooler air up high associated with a trough or low aloft drifting over the area, then enough instability could be present to let things get started. Also, it seems likely that some mild and perhaps localized surface trough or low, not enough to show on a chart, would have to be present in order to start lifting for nocturnal storms.

However they form, nighttime thunderstorms can be as mean as the daytime variety. Even though they often have higher bases, somewhat like the storms on a warm frontal slope, all the bad things are there.

At night the lightning is always visible and is a good warning sign of storms. But most pilots with a lot of night flying experience in thunderstorm country will suggest that this way of avoiding storms is only a fairly approximate technique. As a result, if there is a cluster of storms, the visible lightning would be useful only in avoiding the whole cluster, not in working your way between cells.

ONCE UNDER WAY

Regardless of what spawns it, the structure and effect of a thunderstorm is the same. The moist air feeding the storm is drawn predominantly from the lower levels, pulled in by the updraft that develops as the relatively warm cloud accelerates into cooler surroundings aloft. This influx of air causes subsidence around the cloud, something that every pilot has noticed when trying to climb in the clear areas between developing cumulus. There's a settling effect between the clouds as the warm and moist air is pulled in for the big trip aloft. It's rather like having your air pulled from under you, if you'll pardon the terrible pun.

When the storm matures and rain starts falling, the general area of the storm becomes turbulent because the falling rain brings with it a downdraft, while the updrafts still continue around the outside of the cell. The interaction between the upward and downward moving air results in spectacular disturbances. These can be found a considerable distance from the rain or from cloud, especially at lower altitudes, so storms need to be given a wide berth. (As a matter of fact, most pilots even avoid building cumulus clouds by a good margin because these fat little fellows can dish out a brief but severe pummelling.)

FORECAST

When the forecaster includes the chance of thunderstorms on the day's fare, he or she is telling us that the ingredients might be there. The moisture supply is perhaps the easiest of all to predict. The flow in the lower levels and the surface moisture is the tipoff on that. Lifting action is fairly easy to predict on a broad-scale basis, too. But a front, for example, can develop, dissipate, speed up, or slow down and modify the action or the location of the lifting effect. This would modify the chances of formation or induce a positional error. Instability is probably the greatest source of uncertainty. The temperature aloft is measured, but as troughs and ridges move across at the 500 mb level, the situation in the real breeding ground of cumulonimbus can change rapidly. And if the building cloud can't get support at the 18,000 foot level, it's just not going to amount to a lot. Remember that temperature drop of 2 degrees C per 1,000 feet; it is critical because that is the usual rate at which moist air, that in a cloud, cools with altitude. If the environment in which the cloud is trying to grow is cooling more slowly than that, storms shouldn't develop. If the air around the cloud is cooling more rapidly, then there will be support for vertical development.

When the forecaster goes past listing a "chance" of storms, conditions look more ripe for development. The forecast of "occasional" activity means conditions will be conducive to thunderstorm formation at least sometime during the period covered by the forecast. The flat forecast of thunderstorms is an indication of as much certainty as a weather forecaster can muster. A convective Sigmet is issued when a probability of strong storms exists, when they develop and are not covered by a forecast, or when they are in progress.

There will be times when storms don't occur despite an enthusiastic forecast of their likelihood, but we had best not use such events as an excuse to view all thunderstorm forecasting as crying wolf. Instead, use missed forecasts as a reminder that weather is what you find, not what is forecast. It works both ways, too: thunderstorms can occur when none are forecast, though that is a less likely event than the opposite situation.

There is good information available on thunderstorm possibilities. A good start in the morning is the "Today" program weather coverage. They show the areas where there is a possibility of severe storms, as covered by the National Weather Service analysis. This is no guarantee that something will develop, but there will probably be enough instability to get things going. There's also aviation weather on PBS television stations in many areas.

Any indication of the possibility of severe storms should be compared with the synopsis to get some feel for what is causing the necessary lifting. Is there a surface low to the west or southwest of your position, or a cold front to the west? If you can get a look at, or a description of, the 500 mb chart, is there a low or trough to the west? If so, is the jet stream making a trip around the south side of the low or trough? That makes for the worst possible situation, and remember that the squall line in that case usually forms well ahead of the advancing cold front. Or is cold air being moved from the north or northwest down over very warm air at the surface? Just examining a forecast of severe storms out of context does little good. There is always a triggering mechanism and we must know what it is in order to make our plans.

Weather sequence reports can help a little, especially after storms begin developing. A thunderstorm is reported at a station whenever thunder can be heard and, if more than one station in a general area is reporting a thunderstorm, you might well surmise that activity is widespread. The report of towering cumulus or cumulonimbus in the remarks section of a sequence report is also important.

RADAR

A better source of information is weather radar. Information from radar comes in various forms, a widely used one of which is the radar summary chart. This shows precipitation areas and pinpoints thunderstorm cell activity. It also shows maximum cloud tops, direction of cell movement, and direction of area movement, as well as whether the activity is increasing, staying the same, or decreasing in intensity. The information is not brand-new—it is very important to note the time of the observations used to compile the chart—but it is super-useful in telling what precipitation patterns are being formed by a general weather situation.

The information on movement can be used to project locations to the present time, and the information on intensity can be used to get some idea of whether it is better or worse now than at the time of observation. Any indication of an increase in intensity bears a special importance because it means conditions are good for development; not only might the existing storms strengthen, but others might form and extend the area of coverage. This is especially true where a line of thunderstorms is involved. If there is a relatively short line, increasing in intensity, it's wise to mentally extend that line in both directions and be wary of development that will lengthen the line.

Even storms that are associated with some gold-plated triggering mechanism, such as a deep low or jet stream effect, have their ups and downs during the day. Morning, when there has been no lifting from heating, is usually the calmest time. I shall always remember one especially strong line that spawned tornados across Arkansas, Mississippi, and Alabama. It had done the dirty deed in northeastern Arkansas just before midnight the night before, and I was examining it from central Mississippi about ten the next morning. It didn't look bad on radar or visually, and I was able to fly underneath the line with hardly a bump and only some scattered heavy showers visible to either side. But by that afternoon it had rejuvenated and it proceeded to scatter several small towns in Alabama across the landscape. So when examining radar reports, or when looking at an actual radar scope, take careful note of the time of day. Things might not be so bad now, but they could be a lot worse by the time the area of weather is reached. Also, when you are in an area where it's visually apparent that cumulus are building rapidly, don't be surprised when a thunderstorm shows up where a few minutes before there wasn't one.

CHARACTERISTICS OF RADAR

Radar is the most commonly used thunderstorm detection device, and the basic characteristics of this tool are worth understanding even if you don't have the equipment in your airplane.

Radar shows precipitation, and the nature of precipitation in a thunderstorm is used to differentiate between an area of plain old heavy rain and a real live storm. Rainfall rate (seen as reflectivity by radar), and the change in rainfall rate, is the prime tipoff. If there is an area of very heavy rain surrounded by areas of light or no rain, and the demarcation line is a thin one, then you can bet that the heavy rain is in connection with convective, or thunderstorm, activity. When thunderstorms become embedded in rain, or when the radar is looking through rain toward an area of thunderstorms, then the picture is not so clear and it can be difficult to pinpoint the cells. But in any case, ground-based radar can be used to measure cumulonimbus tops and give a clearer picture of where the most severe thunderstorms are located. The taller a cumulonimbus, the greater the velocities likely found in both the up- and downdraft, and the greater turbulence in the areas between upward and downward action. So height is a very important parameter.

Radar comes in three variations: weather radar, airborne weather radar, and air traffic control radar. Regardless of which one is used, consider the information as giving only the location of precipitation. The radar return does not itself tell us the location of turbulence. It's not the cell alone that we must avoid, it is the area around the cell. The greatest level of turbulence is often not found right in the area of strongest radar return. Remember, where the rain is falling is where the downdraft is located; the downdraft can be rather smooth, at least on a relative basis. Stronger turbulence can be found where there is interaction between the downdraft and the updraft or between the downdraft fanning out over the surface and air being drawn into the storm. Especially in lower levels, severe turbulence can be found quite a distance from a storm as the downdraft turns and becomes a strong surface wind beneath the air rushing into the storm, moving in the opposite direction. See Figure 15. The classic roll cloud that forms ahead of very active storms or squall lines is characteristic of a most enthusiastic interaction between outflow and inflow. The air literally rolls over there.

If you see a gap in a line of storms when looking at radar, and the gap is only a few miles wide, place no bets on flying smoothly through it. The same goes for information on a "light spot" given by a controller. It's better to have a lot of space in moving through any collection of storms, whether it be a line or a cluster. Guidelines based on accident history suggest at low altitude a margin of five miles from the nearest detectable

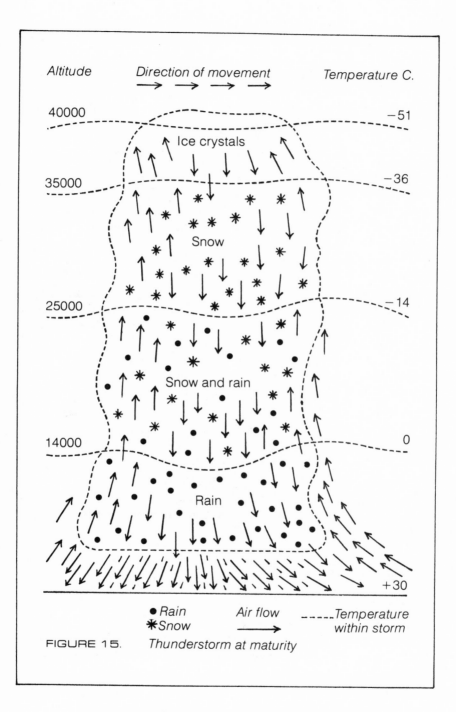

Altitude	Direction of movement	Temperature C.

40000   −51

Ice crystals

35000   −36

Snow

25000   −14

Snow and rain

14000   0

Rain

+30

● Rain   Air flow   - - - - - Temperature
✳ Snow   ────▶   within storm

FIGURE 15.   Thunderstorm at maturity

cell, increasing to twenty miles when severe thunderstorms are forecast.

When you are in an area of storms, what you see can count for as much as what the radar sees. The rate at which the cumulus are building, and a visual assessment of the height to which the clouds have built, can often tell a better story of what to expect in a developing situation than radar can. When moving through an area of rain at a fairly low altitude, the darkness of areas tells a lot about rainfall rates—the darker it is, the heavier the rain is likely to be. What you see is no substitute for radar, but it should never be completely disregarded in favor of information from an electronic device.

A device that senses and plots lightning strokes, the Ryan Stormscope, has been developed and marketed as a weather avoidance device. While it operates on a completely different principle from radar, it is certainly worthy of consideration. Lightning strokes are plotted on a cathode ray tube, with an accurate azimuth portrayal and an approximate distance portrayal. Distance is approximate because the positioning of the strokes on the screen (in the form of dots) is based on the strength of an average lightning stroke. One stronger than average will appear closer, one weaker will appear farther away.

One thing is certain: When there is lightning, the possibility of turbulence is very high. Thus the Stormscope is an excellent device in warning of this type of turbulence. Like radar, it must be used with guidelines for avoidance. In the case of the Stormscope, the primary thing is to not fly toward any area where lightning activity is displayed when the range setting on the device is on the closest setting.

NO GUARANTEE

With any weather avoidance system, there is no guarantee of a smooth ride even when all guidelines are followed. When conditions are such that thunderstorms are building, the instability of the atmosphere can result in general turbulence, as well as turbulence in areas other than those where lightning and heavy rain-producing thunderstorms are present. What the electronic devices can tell you is where there is almost certain to be severe turbulence. And as has been learned the hard way many times, whatever we use—vision, radar, or a Stormscope—should be used for avoidance, not for penetration of a thunderstorm or of a closely spaced cluster or line of storms.

When moving around a storm, some rules-of-thumb suggest that the most severe turbulence is often found on the side toward which the storm is moving, and the side from which it is feeding. Turbulence is also likely to be found a greater distance away from the storm in these areas. Storm cells usually move to the east or northeast and are fed by low level southerly or southeasterly winds, so that gives some clue to areas where the fury of the storm is likely to be stronger and more widespread. Turbulence worthy of the name can be found on all sides, though, and it's never advisable to cuddle up close to a storm when circumnavigating it. At times we can see areas of turbulence around the edges of storms, as defined by cloud wisps, but at times the turbulence exists in an area where it's absolutely clear.

SHEARED

Wind shear is a term used to describe a rapid change in wind direction or velocity with altitude or over a relatively short distance. There is wind shear around thunderstorms as the wind flow characteristics of the storm interact with the general weather situation. There's also wind shear when an active front is moving through an area, or when you are flying at or near the slope of a front. The turbulence associated with up and downdrafts around mountains can be considered a form of wind shear. At night, when the wind becomes calmer near the surface because of cooling and the development of an inversion, and the wind aloft remains at its daytime value, there is wind shear at the level of the inversion. When flying down between the mountains, the turbulence we feel in addition to the ups and downs is wind shear as velocities ebb and flow while the wind moves through the irregular terrain.

Wind shear can have a pronounced effect on an airplane. Either flying into a rapidly increasing headwind or with a rapidly decreasing tailwind causes a temporary increase in airspeed. Or if airspeed is held constant it feels like an updraft. Flying into a decreasing headwind or with an increasing tailwind causes the airspeed to decay temporarily. Or if the airspeed remains constant it feels like a downdraft. If the change is very gradual, you may hardly see or feel it; if the change is rapid, it'll sure show up and it can affect the safety of flight operations.

WHAT HAPPENS

It's pretty easy to visualize what the various situations can cause. For example, if you are low and slow on a final and encounter a rapidly decreasing headwind or rapidly increasing tailwind, the arrival might well be at a point other than that of your choosing because of a rapid decay in airspeed.

If you are flying at low altitude toward a storm, from the side toward which it is moving, you'll fly from the area influenced by the inflow into the storm (tailwind) into the storm itself (or into the interaction between up- and downdraft) and the result can be a rather spectacular increase in airspeed. Just as you get that in hand, you could move into the downdraft and an equally spectacular sinking spell.

The important thing is to know enough about small-scale and localized weather influences to anticipate wind shear. If there is a thunderstorm nearby when landing or takeoff is contemplated, this means evaluating the proximity of the storm and delaying the landing or takeoff (or going to another airport) if there's any chance that the influence of the storm will move across the airport, or across the arrival or departure path while you are using it.

If engaged in the extremely risky business of landing or taking off when a thunderstorm is affecting the airport, the only salvation might be found in an accurate assessment of the influence of the storm on the path to be flown. If taking off away from the storm, for example, the chances are an increasing tailwind would be prevalent in the first stages of climb, making climb performance rather sad. If taking off toward a storm, climb performance might be good to begin with, but it might suffer terribly as the airplane enters the heavy rain, which is related to the core of the downdraft. And regardless of direction, wind shear turbulence might create very genuine aircraft handling problems, especially in the takeoff or approach configuration. It is *very risky* to take off or land with a thunderstorm close by, and it is even riskier if done without a good understanding of the dynamics of the storm.

PLOT AGAINST IT

It's relatively easy to prepare for wind shear in less critical situations just by using information that is readily available to you in flight. For example, if you are flying northeastbound with a strong tailwind, and you find that the ILS approach will be to Runway 4, surface wind calm or easterly at some low value, you *know* there will have to be wind shear on the approach. Where depends on several factors. Such a situation is indicative of a cool front that slid under a strong southwesterly flow and stopped. There is an inversion (warmer air aloft) in such a case, and there's a good chance of low level clouds at the base of the inversion. So the shear might well be encountered as you enter the clouds at a thousand or so feet above the ground. This can harass you on the ILS approach. Here you come, flying in nice smooth air, needles parked in the proper places. Then, at or just inside the marker, you enter clouds *and* turbulence while at the same time encountering an increase in airspeed and corresponding resistance to descend on the part of the airplane. (*Remember,* when first encountering a decreasing tailwind the effect will be to increase airspeed.) If this is anticipated, it will be a lot less trouble.

ANALYZE TURBULENCE

The nature of turbulence tells us a lot about the cause of turbulence, and it's wise to analyze turbulence when it is encountered to get some idea of what might come next.

Convective turbulence, that associated with the updrafts making cumulus or cumulonimbus, has a definite upward trend. It's experienced in the airplane as climb. If the power and airspeed remain constant, the altitude will increase. The upward vertical speed is related to the strength of the updraft. There can be a pronounced increase in airspeed when first encountering a strong updraft. When we get to the turbulence associated with the downdraft, everything reverses; it is all down. Upward motion is associated with something building and the downward with the subsiding. The same up- and downdraft situations can be found over and on the lee side of mountains when the wind is strong. In case of the thunderstorm, an updraft means it is still feeding. Over the mountains, the updraft means cloud development might occur where the updraft is found. Icing could be a big problem here because of the rapid lifting of warmer and more moist air into cold air aloft. The downdraft is just the opposite.

Contrast the up and down turbulence with straight wind shear turbulence, which is often found between up- and downdrafts but which can also affect large areas where there aren't any up- or downdrafts. Wind shear is usually comprised of short jabs. Bang, bang, bang. The airspeed fluctuates, wildly at times, but there is no continuous well-defined updraft or downdraft action. The airplane can be difficult to fly as it wallows, rolls, and pitches in response to the turbulence of air. Away from storms, and up high, the most severe wind shear turbulence is encountered around the edges of the jet stream. For low altitude pilots, it's along frontal slopes and in unstable and gusty wind conditions.

Remember how shallow a frontal slope can be? A 1:100 example was used for a warm front, meaning that the slope would be about 5,000 feet above the ground 100 miles ahead of the surface position of the front. So if you started getting knocked around 100 or so miles ahead of the position of the surface front, shear would be a likely cause. If the turbulence consisted mainly of sharp jabs, without the up and down action, it would tend to verify this. If a smoother ride were desired, climb or descent might be the ticket.

Knowing the positions of fronts is important if you are to come up with a best bet on whether to climb or descend to minimize shear turbulence. For example, if you were flying north away from a warm front, climbing might get the airplane above the slope for the time being—but you'd fly back into it after a while. And if you were flying toward a warm front from the north, a descent would tend to correspond with the slope of the front. You might go below it but fly back into it later. Other factors enter into the question of which is better, climb or descent, and all should be considered. Moving above a warm frontal slope would tend to increase the chances of encountering convective turbulence, because thunderstorms usually build from the slope upward and general instability would be stronger up there even without thunderstorms. A descent into cooler air below might increase the chances of encountering ice, especially if conditions were good for freezing rain. All factors have to be weighed. Whatever you decide to do, a mental picture of the frontal slope and the temperature distribution above and below will help you change altitude in the best direction.

In convective turbulence, an altitude change isn't likely to bear much fruit because the disturbance probably covers a lot more vertical territory than does shear turbulence. Also, changing altitude can be very difficult. If you are in a strong updraft and decide that you'd be more comfortable down lower, there might not be much choice. The airplane might not be capable of descending at any reasonable speed even with power off and the landing gear down. When it's convective turbulence, things go better if you get the altitude correct before reaching the turbulence, and many feel that the best altitude (other than one comfortably on top of all clouds) is the lowest possible safe altitude—below the cloud bases if possible.

When learning about weather from turbulence, do be aware that it can change from one variety to another rather quickly. For example, if you were penetrating a textbook thunderstorm you might encounter wind shear turbulence, then updraft, then shear, then downdraft, then shear, then updraft, and, if you are still there, a final bout with shear turbulence. If you were in a jet with high wing loading it might feel like a relatively brief tussle with the elements; in a light airplane it can be more like a 15-rounder.

HELPFUL HINT

There's a nice little rule-of-thumb that relates to wind shear that is always worth repeating. If you are climbing and encounter a substantial increase in climb while maintaining airspeed, and there is no suggestion of updraft caused by convective activity or wind over mountains, the bonus climb is probably a result of an increasing headwind. It usually comes at a level where the turbulence starts to go away, and where any clouds start topping out. What it tells you is that if you opt for the smooth ride on top, you get a lot more headwind than you would on the bumpy ride below. Such is life. On the other hand, if you have trouble coaxing the airplane upward as you move into the smooth air aloft, it could be a sign of a rapidly increasing tailwind and an offer of the best of all worlds—smooth air and a higher groundspeed.

THE LOOK AND FEEL

The look and feel of clouds and precipitation tells us a lot about a weather system. The air is smooth in stratus clouds, which are a product of change that is often relatively gradual, and a stable situation. It is bumpy in cumulus clouds, a product of an unstable situation. A large area of steady rain means the air is relatively stable. Showery conditions mean the air is unstable. Big drops (or snowflakes) mean it is unstable.

As we pick around in frontal zones, it's important to get as much information as possible about rain because of its relationship to turbulence. It's also important to pay attention to what we fly through. If the rain is steady, and goes on for quite a while, the ride could be a pretty good one. If rain is showery, and if the flight moves through various cloud formations with vertical development of cumulus quite evident, then the ride might not be so good. In the latter case, it becomes critical to learn as much as possible about the extent of development in the area. Are you flying toward the low or front (toward the likelihood of greater instability), or away from the low or front? What does the traffic controller's radar show? Is there static from thunderstorms evident on the low-frequency ADF receiver? When flying through a shower, how large are the rain-drops? The more vertical action, the bigger they get.

The possibility of embedded thunderstorms always raises questions about the wisdom of low altitude IFR flying. These thunderstorms would most likely come east of a low center and north of a warm front if one is depicted. They can't be embedded from base to top, but if you were operating in a general cloud layer, you could move close to the storm without seeing it. Usually, though, the cloud layers are not too thick, and a pilot can move to a between-layers condition in any area of embedded storms. That doesn't mean it will be possible to get through visually, but it could mean that you'd be forewarned of bad things by the dark appearance of the sky ahead. Even when summer storms are embedded in thick low-altitude haze and smog, their darkness is usually visible well before the turbulence is reached. Airplanes seldom venture into disastrous turbulence without the pilot seeing or feeling something in advance that suggests trouble.

The relationship of precipitation to the forecast offers some suggestions, too. If no rain is forecast but it is raining heavily over a wide area, a large-scale ascent has developed that wasn't considered in making the forecast. There is probably a brand-new low pressure around. If the unforecast rain is showery rather than general, the air is unstable. Likewise, if the forecast was for rain but there's nothing but drizzle for miles around, things are better than anticipated. With light rain or drizzle, tops are probably lower, or at least the clouds are layered. Heavier rain means higher tops, usually above the ceiling of a piston airplane without turbocharging. Showery conditions suggest high and uneven tops.

ICE

All this business about clouds and precipitation relates very much to airframe ice. This nasty little phenomenon is made possible by supercooled water droplets, that is, water droplets that remain liquid even though the temperature around them is below freezing. When we fly into them, they splatter and freeze on the airplane. Bad business, and something to avoid.

The nature of clouds has a lot to do with icing. Stratus clouds offer rime ice, a rough textured collection that forms when the tiny water droplets of stratus are busted. Cumulus clouds offer clear ice that smoothly coats the front of the airplane and the propeller when the larger water droplets of cumulus are disturbed.

Stratus icing is relatively easy to plot against because the vertical extent of stratus is not great. Climb or descend to get out of the ice. Too, when stratus clouds are old, there might not be any ice because the small droplets will finally freeze. Or if the temperature is a great deal below freezing—say -15 C—the clouds are probably ice crystals. Freezing rain also comes in smooth air, and we covered that when discussing warm fronts.

Unstable conditions and cumulus clouds give irregular doses of ice just as they are related to showery conditions. The vertical extent of icing in cumulus clouds is likely to be much greater than is found in stratus clouds. For one thing, the clouds are taller. For another, the vertical action and mixing makes big drops out of little drops, and the bigger drops can remain supercooled at much lower temperatures.

Where do we find cumulus clouds in the wintertime? There can be a lot of them in warm frontal zones, in instability behind cold fronts (colder air flowing over a warmer surface), over mountains, and in cold frontal zones. A logical area for icing is in or north of the path of a developing low in the wintertime, where there is maximum mixing of moisture and cold air.

LIGHT AND SHADE

Light can play tricks, and late in the afternoon the sky can look fearful when it really isn't so bad. But most of the time the various shades of gray tell a real tale. The darker the gray, the worse the deal. Dark is usually associated with heavier rain, and with extremely heavy rain the view ahead can even take on a gray-green appearance.

When cumulus are building rapidly, they might well be a lot bumpier than cumulus with lackadaisical growth.

The look and feel of clouds and precipitation has to be

considered when comparing conditions with what was expected. It's a cinch that a pilot who continues in accordance with the original plan when it feels and looks worse than anticipated is likely to have misadventures with weather—more than the pilot who completely reevaluates the situation anytime things aren't according to plan.

BASIC EXCEPTIONS

We often start off with the impression that highs are good, lows are bad, and fronts are brick walls in the sky. Experience teaches us that the weather can be grungy when the pressure is high, it can be nice when the pressure is low, and while some fronts might be brick walls, others are pussycats. And we can find weather that looks and feels like a front when there is no line drawn on any map.

There is always a perfectly good reason for any variation from the basics. For example, we know that air assumes the properties of the surface over which it flows, and if a high pressure area is bringing an onshore flow, moving warmer air over colder ground, zap, you can have very low ceilings and visibilities in fog. The ground, and the air next to the ground, is cold and moving the warm moist air over it creates an inversion virtually at the surface, or at least at a very low level. This traps the air at the surface and results in persistent fog. The high might be a shallow one, influencing only the surface flow. A strong southwesterly or westerly wind aloft can add to the creation of the inversion. This might be considered small-scale weather, at least in a vertical sense, because it affects conditions only very close to the surface. But it can be widespread in a horizontal sense, and can have quite an impact on flight operations.

Conversely, the weather can be pretty good even in the "bad" sector of a low. Even with a surface low on the chart, a lack of upper level support can preclude much development into the upper atmosphere. Or a weaker low might not have strong enough circulation to pull in the moisture necessary for widespread cloudiness or precipitation. An identifiable front can have little weather and not enough wind to worry over, where a trough that never really makes it to the weather map can contain elements of weather that seem quite significant, especially to a pilot flying a light airplane at low altitude.

We deal with these seeming exceptions to the rule by staying alert and paying more attention to what we *actually* see and feel than to what we *expect* to see and feel in the way of weather. Remember, meteorology isn't that precise, and most of the information we get is based on observations made at the surface or from satellites far above the earth. Nothing but the airplane takes the measure of the weather at the chosen flight level, along the chosen route of flight.

Interpreting
Weather Data

There are a lot of pitfalls as we search for weather information, and the primary one is that the information system seems more in tune with the need of ground than airborne users. If the time of frontal passage is missed by a few hours, it doesn't matter to most housewives. For a pilot, however, a two-hour miss can be critical. It can mean the difference between zero-zero and good VFR at a destination or at an alternate.

Delay is built into our weather information system, and at times it seems almost to flaunt a lack of precision. For example, I was keeping up with the weather one afternoon because some strong storms were supposed to move through, and on the evening news one of the leading New York TV weathermen said that there were indeed big storms that very minute just to the west of Trenton, New Jersey. I happened to be east of Trenton, New Jersey, and it was absolutely sparkling clear to the west. The storms had moved by an hour earlier. To a ground person, perhaps that wouldn't mean too much. If the aviation weather were as far behind, it might mean a lot to a pilot planning to land in the area.

In researching this book, I found a lot of discrepancies in the information that is disseminated by the FAA. I'd dutifully re-cord a morning weather briefing, including the synopsis described by the briefer on a map, only to find later that the forecasts and the synopsis were very much in error. The National Weather Service offers for sale daily weather maps, issued weekly to cover the Monday through Sunday period just past, and these include the 7 A.M. surface map, the 500 mb chart, temperature, and a precipitation map. The synopsis described in a morning briefing from the FAA often just isn't like the one on the official NWS map for that morning. The trouble is, you can't buy the real map from the NWS until well after the fact, so we have to depend on a combination of the information we get from the FAA plus our own ability to adjust for the time lag and for any inaccuracies. This isn't difficult—you have only to recognize that it must be done and study the subject. In many instances, as you fly along and examine the true picture, you can draw a better weather map than is available from any source.

This isn't to ridicule the National Weather Service or the FAA. Their people are dedicated and their information is the best available. It's just that any government agency eventually comes to serve primarily itself, and the resulting lethargy is bound to lag behind something as dynamic as weather. Combine another dynamic factor—the airplane—with weather, and it becomes even clearer that pilots must have weather interpretive abilities. Never fail to get every bit of available information, but never fail to question its timeliness and accuracy.

Pilots flying relatively slow airplanes—those cruising at or below about 200 knots—have more opportunity to interpret weather than pilots flying fast airplanes. While you often hear of a critical need for better cockpit information for air carrier crews, the critical need in slower airplanes is for pilots to take advantage of all the available information. There is more time to look at and experience the weather, the airplanes move into changing conditions more slowly, the fuel supply is often less critical, and the flight is more often conducted down in the weather instead of in that usually serene air above everything. When flying low and slow, there is no excuse for not keeping up a minute-by-minute interpretation of the unfolding weather picture.

Almost all our information is based on surface observations, and observations above the levels most often used by light airplanes (including satellite observations). However, this system has withstood the test of time because the surface weather gives us information on what is happening in the very low levels, near the ground, and the upper air observations tell us what's going on up there. All we need is the knowledge to connect the two.

There are small as well as general peculiarities in the weather information system. For example, it took me years to learn that round weather radar scopes can lead you down a primrose path. More than once, I talked to a person looking at a scope, heard that person say that it looked like a line of storms ended somewhere, and then went flying off toward that somewhere only to find that the line didn't end there. It just ended there on the round scope. Looking for the end of the line was like looking for the pot of gold at the end of the rainbow.

Another lesson learned with time is that forecasts of winds aloft are notoriously inaccurate at lower levels. I've found that by getting winds for both the current period and for the next time period, it's possible to anticipate errors.

In fact, when you perceive that systems are moving a bit ahead of schedule, the winds for the next period are often more applicable than those for the current period. If they are moving slower, the reverse might be true.

One reason the wind forecasts are so bad is that the criteria for evaluating them are extremely broad. If the forecast wind speed is 25 knots or less, the direction must be 45 degrees in error to mandate a change in the forecast. If the velocity is over 25 knots, the criteria is 30 degrees. When the wind is forecast at 25 knots or less, a 10 knot error is allowable; it's 15 knots for forecast velocities over 25 knots. The forecasts for temperature aloft are allowed a 5 degree C error, except at 6,000 feet where a 3 degree C error dictates an amended forecast.

All that probably sounds very reasonable to the National Weather Service, but consider what it can do to a 120 knot airplane. For example, with an actual west wind at 39 knots and a forecast wind of 240 degrees at 25 knots, a northwestbound pilot would find his or her groundspeed to be 89 knots instead of 111 knots, as calculated before the flight based on the forecast. In such a situation, a pilot would probably have to replan the entire flight and go to an alternate course of action, especially if the proposed flight was a long one. And despite the fact that the pilot would feel that the wind forecast was grossly in error, the NWS would judge it accurate and not revise it. In the same way, a big leeway in temperature forecasts can mean a lot to a pilot if the temperature is close to freezing and ice is the question of the moment.

There is a bright side to all this. Once a pilot understands the situation, flying can become both more interesting and safer. Instead of pressing on in blissful ignorance, taking everything at face value, the pilot can enjoy the challenge of figuring out what is really happening. The flight can be conducted with the emphasis on understanding the weather and trying to minimize the effects of weather. Weather becomes primary; reduction in the distance to the destination remains important but can become secondary, at least to the extent that the pilot avoids pressing on at all costs. The flight winds up with both educational and transportation value.

Understanding the three-dimensional nature of weather puts us one up on most folks, too, by removing much of the mystery that surrounds the subject. The fact that we fly helps us understand things like the large-scale ascent of air ahead of an advancing low center. We can experience frontal slopes and can feel wind shifts.

Experience is the key to understanding, and after each flight we can turn every fact, everything we saw and felt, and every bit of information over and over in mind, to examine it thoroughly and learn as much as possible from it. The pilot who walks away from the airplane without reliving the flight and the weather, and the total experience, sure isn't getting full value.

The next four chapters of this book cover actual flights and their relationship to weather. Each chapter covers a quarter of a year. And as we move into the experience section of this book, ponder for a moment that this is all hindsight. These tales of actual flights were written after the airplane was safely on the ground. The maps were drawn based on the synopsis described during the briefing before the flight as well as on the official NWS charts published later, and the two don't always agree. All this might suggest Monday morning quarterbacking to the extreme, but there's really no other way to learn about weather. There's no such thing as figuring it all out in advance. And once we've flown, we need to study. Some situations are straightforward and the explanations basically simple. Others are complex, seemingly at odds with the basics of meteorology. But there can always be a logical conclusion. The only mystery in weather comes when it is examined with a closed mind, in an attempt to make it conform to what we thought it would be instead of what it really is or was.

Most of the flights described were flown in a Cessna Cardinal RG, a single-engine retractable. This airplane is not turbocharged for high altitude performance, so the flights were generally flown below 12,000 feet. This is not the best part of the airspace from the standpoint of staying out of weather, because a high percentage of the clouds that roam over our planet are at these lower altitudes. But it is a *good* place to fly if you are learning about weather; you fly through instead of over the weather that develops.

A device called the Ryan Stormscope is mentioned in relation to some thunderstorm situations. As explained earlier, this equipment plots lightning strokes on a cathode ray tube. It was installed in the airplane for most of the time during which information was gathered for this book.

Most of the flights were flown under instrument flight rules because it is the author's feeling that when there is enough weather to talk about, risks are best managed by flying IFR. All flights were flown in the central and eastern U. S.

Flights: First Quarter
January, February, March

In the first three months of the year, we have the most blizzards. And, later in the period, the sometimes nice and sometimes severe weather of spring begins in the southern states. Weather systems tend to move rapidly because the steering influences of the upper air patterns are strong. Lows that form in the Gulf of Mexico and off Cape Hatteras have a definite effect on weather over a wide area and strong storm systems are a factor over the Great Plains. Thunderstorms can be a problem in the south and southeast, especially toward the end of the quarter.

Icing can be a big consideration, especially to the north of low pressure centers, in waves on stationary fronts, and ahead of warm fronts. Strong winds over mountains can cause turbulence up to altitudes well above the ceiling of normally-aspirated airplanes, and strong winds everywhere can create agonizing headwinds (or, happily, tailwinds that contribute to memorable flights).

The best thing that can be said for the weather in this period is that it changes rapidly. An exception can be found in the south and southeast, especially in January and February, when cold fronts often become stationary and create widespread areas of bad weather that seemingly last forever. But eventually a low will form on the front and whisk all the bad stuff away.

Icing, and the difficulty of starting cold engines, are probably the most frustrating factors, especially in January. Despite all the drawbacks, it is still a very flyable time of year. It just takes patience.

JANUARY—TRENTON TO VERO BEACH

Warm Front, then a Cold One

The FSS briefer said there was a low pressure center in eastern Tennessee, a warm front across northern Virginia, and a cold front to the southwest of the low. The winds aloft were very strong from the southwest. Ceilings were appropriately cruddy—all down around 500 feet—and it was difficult to find a legal IFR alternate for the southbound trip. The problem was compounded by the strong wind. I finally decided on four hours and fifteen minutes to Wilmington, N. C., with Myrtle Beach as an alternate. For arrival, Wilmington was forecast to be 500 overcast and one mile visibility with a chance of a thunderstorm. The radar summary chart showed a scattered to broken area of rainshowers along the route of flight with maximum tops at 23,000 feet. It was relatively warm on the surface at Trenton and the southwesterly flow aloft suggested that there wouldn't be any icing problem.

After a midday takeoff it was obvious that I was north of the warm front. Rain was falling, heavy at times, and the basic situation at 6,000 feet was one of higher clouds, lower scattered, and occasional cumulus types building through 6,000 and poking up into the higher clouds. There was no electrical activity, as verified by the Stormscope, and there was only some rather mild turbulence in the cumulus. The big news was the wind, which was calculated at 56 knots instead of the forecast 42 knots. Does this happen every time *you* head south in the wintertime?

It seemed a clear-cut warm front and I thought I'd be able to pinpoint a surface position of the front. This was elusive, though. And where the original briefer had positioned the front across northern Virginia, a call to an FSS while en route revealed that the warm front was now shown to be across

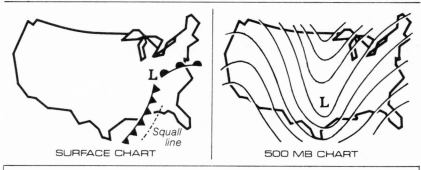

SURFACE CHART | 500 MB CHART

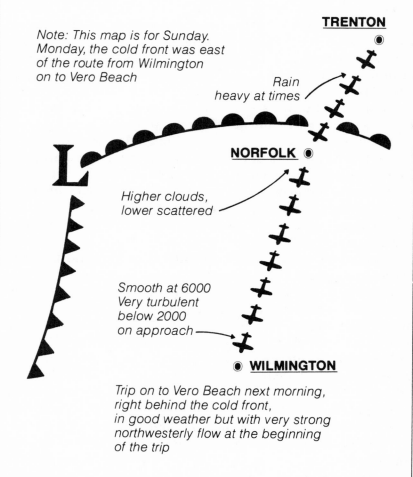

TRENTON

Note: This map is for Sunday.
Monday, the cold front was east
of the route from Wilmington
on to Vero Beach

Rain
heavy at times

NORFOLK

Higher clouds,
lower scattered

L

Smooth at 6000
Very turbulent
below 2000
on approach

WILMINGTON

Trip on to Vero Beach next morning,
right behind the cold front,
in good weather but with very strong
northwesterly flow at the beginning
of the trip

central North Carolina, and the low to be in northern Alabama. Apparently the earlier positioning was incorrect, or perhaps the low and front had just not been clearly defined at that time. It's interesting how the synopsis they read to you does not always change in a logical manner when you are paddling through the stuff.

In the vicinity of Norfolk I flew out into an area with high clouds, clear below. It was as if I had flown through the warm front, but the situation south of the front somehow lacked the tropical look of warm sector weather, proving once again that weather comes in various shades of gray. The trip on to Wilmington was uneventful, and the weather there was excellent —much better than forecast. The warm front had definitely passed over, as the surface wind there was very strong from the south and the turbulence below 2,000 feet was quite heavy. The altimeter setting was low and dropping rapidly, suggesting that the cold front would soon be there. I pulled in at Wilmington after dark and found refuge in the local Hilton.

During the night, the low gained a lot of strength and then moved to the north. The cold front passed Wilmington with an extremely strong flow—strong enough to carry freezing temperatures well down into Florida—and there were some severe thunderstorms as the front passed through Florida.

The weather briefing the next morning offered no real problem, although there were plenty of Sigmets for turbulence. Those things appear unnecessary—it's always bumpy down low when the surface wind is 35 knots—but they were duly noted. Interestingly, they didn't mention icing, even though a cloud deck was moving in with the westerly wind and the temperature had plummeted from +16 to +7 C in a very short period and was still dropping. My original filed altitude of 6,000 feet was in the cloud tops, where there was some light rime ice accumulation, and I had to go up to 8,000 feet to get on top. The cloud deck was typical of what follows cold fronts —especially when the temperature drops precipitously and the cold air over very warm ground results in instability.

Wind is weather, and on this flight I encountered wind aloft as significant as I have ever seen. From the long sheet and from the controller's observation of groundspeed, the leg from Charleston to Savannah was at 95 knots on a magnetic track of

233 degrees. After Savannah, the track was 196 degrees and the groundspeed increased to 143 knots, calculated as well as recorded on the traffic controller's scope. The forecast was for the wind to be from 300 degrees at 47 knots. If the direction was correct, the velocity had to be much higher than 47 knots for the groundspeed to have worked out as it did. The low in this instance was a very strong one, as was the front, and it was an example of how winds can be stronger in the immediate vicinity of a front than forecast, stronger perhaps than 100 miles behind the front.

JANUARY—SPRINGFIELD TO JOPLIN

An Icy Evening

The day started in New Jersey and I reached Springfield, Missouri, at sundown after flying all day in good weather. The last part of the leg was flown at 8,000, IFR, and there was just a trace of ice in the clouds. The flow was southwesterly at 25 knots, not particularly strong, but I was suspicious of the proposed route on into Oklahoma City, if only because it would be flown at night.

In the FSS, I found that there was a low pretty far out to the west that was developing into a major storm system. I had a warm front on my map from the briefing for the previous leg, but there wasn't one on the chart for this leg. Reports along my proposed route were good. Joplin was reporting 5,000 overcast and ten miles, Tulsa 2,500 overcast and ten miles, and Oklahoma City 1,400 overcast and seven miles. Tulsa and Oklahoma City had strong southeasterly surface winds, suggesting that they were north of a warm front, or whatever might resemble one in the circulation to the east of the developing low. The forecasts for all points called for occasionally lower conditions in snow. The radar unit located near Springfield showed very little precipitation to the southwest, according to a relatively current report. No mention was made of icing, but the surface temperature of −3 C and the presence of clouds carried rather plain messages on that subject.

The logical thing was to look at the situation from an altitude beneath the clouds, where the chance of icing would be mini-

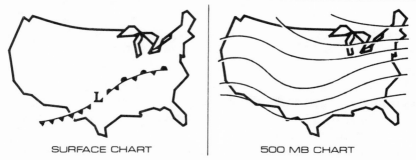

SURFACE CHART | 500 MB CHART

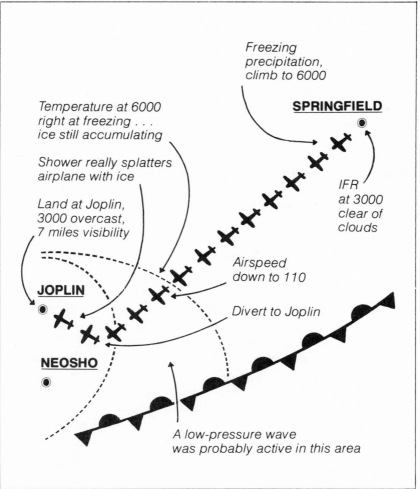

Freezing
precipitation,
climb to 6000

SPRINGFIELD

Temperature at 6000
right at freezing . . .
ice still accumulating

IFR
at 3000
clear of
clouds

Shower really splatters
airplane with ice

Land at Joplin,
3000 overcast,
7 miles visibility

Airspeed
down to 110

JOPLIN

Divert to Joplin

NEOSHO

A low-pressure wave
was probably active in this area

mized. Based on current conditions, the minimum IFR en route altitude, 3,000 feet, would satisfy that requirement as far as Tulsa, where cloud bases should be just over 3,000 feet MSL. The ceiling was lower at Oklahoma City, so that approach wouldn't work all the way. I filed a flight plan to Tulsa, 140 miles away, to have a look. If things were favorable for a continuation, we could fly the other 100 on to Oklahoma City. One at a time. The night had that sort of feel. Examine it a mile at a time and keep one foot on a base.

My passenger caught a little piece of light frozen precipitation as we walked across the ramp, rubbed it between his fingers, and remarked that it was rather sticky. I thought his frozen confection was probably snow or snow grains.

The flying was okay for the first few minutes, then we encountered some light precipitation at 3,000 feet. It was hard to identify, but a flashlight examination of the leading edges and the windshield revealed some ice. That is a mandate for immediate change, so I called the controller and asked if he had any pilot reports. He replied that a twin Cessna near Joplin had reported no ice at 6,000 feet a while back. A clearance to that altitude was granted as soon as I requested it.

The temperature did increase as we climbed to six, and at that level it appeared that it might actually be above freezing. The windshield cleared and remained clear, and in watching the wings with a flashlight it looked as if there might be some melting there. It was hard to tell.

Climb higher? The tops of precipitation were reported at 10,000 feet by radar. Maybe it would be a little warmer at 8,000 but for some reason I doubted it. The temperature had increased from 3,000 to 5,000 and had then remained steady in the climb to 6,000. The slope of a condition similar to a warm front was probably between 5,000 and 6,000; the temperature might drop above 6,000. Go for tops at 10,000? I've not had good luck with radar tops being real tops. For some reason, I didn't want to climb much more anyway because we had gotten quite a sheath of ice on the bottom of the wing in the climb to 6,000. And I doubted that the airplane would have been up to a climb through 10,000 with the ice already accumulated.

Even though it didn't look like more ice was forming, the airspeed indicator was telling a different tale. It dropped from

130 knots to 120, and then to 115. I caught myself increasing power to try and maintain the cruising airspeed. That is surely a temporary patch. Joplin, Missouri, was off to the right, and I was at about the closest point to Joplin when the airspeed dropped to 110. That was a hard message to misunderstand, and the option to continuing appeared to be a landing at Joplin, where there was an ILS and a long runway. Their weather had deteriorated some since the briefing, but it was still holding at 3,000 overcast and seven miles. On request, I was quickly cleared direct to the Joplin outer compass locator and to descend at pilot's discretion.

There seemed no advantage to descending immediately so I stayed at 6,000 feet to have that altitude to parlay into distance should the icing become critical. And it almost did. We flew through a brief but rather hard wet snow or rainshower that really splattered the airplane. It cut a quick five knots in airspeed, and it probably cost more than that because I started a gradual descent right after the shower, having estimated our position to be within a few miles of the airport. In retrospect I guess it was best to have remained at 6,000, because the temperature there was right at freezing and the shower did not overwhelm the defroster and opaque the windshield as it might have in the slightly colder air below.

We broke out of the clouds right over the Joplin airport and I circled to land to the southeast. A bunch of extra airspeed was maintained on final and the sink rate was very high at 1,500 feet per minute with half power and with the gear extended. The airplane was handling okay, though, and the long runway was very adequate for my speedy approach.

I had flown into a similar weather situation once before, quite a number of years before this episode. It was, in fact, the only time I've had more ice on a light airplane than was on my Cardinal RG this evening. That first event was caused by waves forming on a stationary front and moving northeast, contrary to all forecasts. I punched through one of the waves and got very thoroughly frosted when my warm altitude suddenly turned colder.

There was no stationary front shown on the map I saw in the FSS at Springfield, but by the next morning they had one drawn in, just south of the course I was flying, which made the situation similar to that first one. The ice that formed on the airplane was much like the first encounter, too. It was a rough mixture of rime and clear icing, extending far back on the underpart of the wing as well as forming a ridge along the top of the wing, slightly back from the leading edge. In the Cardinal RG, I wasn't leaning forward and looking back at the wing when examining the ice accumulation in flight and I did not see this build-up atop the wing. That ice was why the airspeed was going to pot.

The basic key to the possibility of problems was in the weather map at Springfield. The low was to the southwest, moving northeast. It was forecast to be a bad one, and I was in the sector where the most trouble can occur. There wasn't yet anything aloft to encourage big and bad things in the area where I was flying, but there was enough for me. After the original plan had failed so miserably, landing seemed the best alternative.

JANUARY—JOPLIN TO OKLAHOMA CITY

Snowstorm

One thing is sure about a snowstorm. If you are out on a trip and are shot down by one, today is likely to be as interesting as yesterday. After the diversion into Joplin the night before, on the preceding flight, I knew what to expect when I peered out the motel window at 5:30 the next morning. Snow. Lots of it.

An early call to the FSS brought ominous warnings of icing, too, with a Sigmet for severe icing current in the area. The surface chart now showed a low in the Texas panhandle and a low in west-central Colorado, with that stationary front running eastward. All forecasts called for snow and strong winds. The only hopeful thing that I caught was the fact that the surface winds were all northeasterly. As a low pressure storm system passes, a northeasterly wind is an indication that the storm center is to the south. If it is moving to the east, that means things will get better, perhaps sooner rather than later. If it is moving to the north, it means things will get worse before they get better. I felt

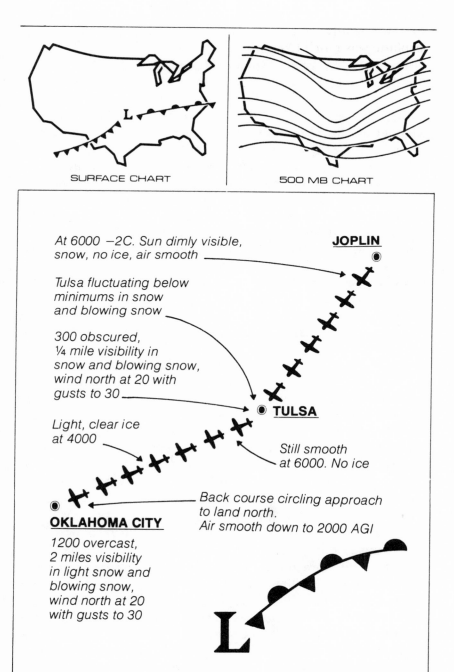

SURFACE CHART

500 MB CHART

At 6000 −2C. Sun dimly visible, snow, no ice, air smooth

JOPLIN

Tulsa fluctuating below minimums in snow and blowing snow

300 obscured, ¼ mile visibility in snow and blowing snow, wind north at 20 with gusts to 30

TULSA

Light, clear ice at 4000

Still smooth at 6000. No ice

Back course circling approach to land north.
Air smooth down to 2000 AGI

OKLAHOMA CITY

1200 overcast, 2 miles visibility in light snow and blowing snow, wind north at 20 with gusts to 30

L

this one was moving east. Brimming with optimism, I went to the airport on schedule.

The problems of an airplane left outside in a snowstorm are not the subject of this discussion, but they are sure a consideration in wintertime flying. The first thing I did at the airport was ask the line person if they could deice the airplane when the weather appeared fit to fly. He said they could, so I went about the business of trying to learn about the weather ahead.

The snow was quite heavy at times, and was playing havoc with surface weather conditions. Some stations were reporting obscurement with a quarter of a mile visibility, and the first airplane to have to go at Joplin missed the approach. But conditions did appear to be improving. By 10 A.M. the sky assumed that lighter look that usually comes as a snowstorm abates, but that Sigmet for severe icing was still in force. The answer to ice seemed to be in pilot reports, and by bugging the FSS to ask for them, and by asking a departing pilot in a turboprop to call one back on Unicom, I was able to compile a dossier of "no icing" pilot reports that I felt invalidated the Sigmet.

An alternate for Oklahoma City was no problem. Amarillo, easily in range to the west, was forecast to clear. Oklahoma City was up and down at the time I filed, but the wind was around to the north and things usually get quite a bit better, at least for IFR, when the wind shifts through north. That means the low is moving to the east, and the cold air being brought down on the backside of the low helps insure that precipitation will be snow, at least in January. Cloud tops are not likely to be high in such a situation.

The air was bumpy on climb but it smoothed out at 3,000. At 6,000 it was −2 C, warmer than on the surface, and I was afraid this might result in some icing but it didn't. The clouds were basically thin and most of the stuff obscuring the sky was snow. In fact, at 6,000 we could often see the ground below and the sun dimly peeping through the snow and perhaps a high thin overcast above. This was all stuff that had made a long trip around the north side of the low, and most of the moisture was gone. Surface conditions were still lousy, though, with Tulsa all but below minimums as we passed there.

Oklahoma City did better for me, and on arrival advertised

1,200 overcast and two miles in snow, wind north at 20 with gusts to 30. After picking up no ice at 6,000, I did accumulate a little clear ice in a lower cloud layer while being vectored for the approach but it was hardly worthy of mention.

The situation this morning was the exact opposite of the night before. Then we had been flying head-on into a developing low pressure storm system. The decisions then had to be based on the assumption that anything bad would only get worse before it started to improve. For the morning flight, the challenge was in deciding when the situation had reversed. In looking at the FSS weather map, with the low positioned well to the west, it looked doubtful. But in examining actual conditions it was apparent that the low was much farther to the east than reflected on the surface chart. And a look at the 500 mb chart showed a pattern that should result in a positioning of the surface low to our south, and a rather rapid movement of that low to the east.

Something else came to mind, too. In the wintertime they seem to be continually locating lows on the chart that never develop. Perhaps there *is* a little surface low in that position, but it just doesn't develop. The thing that we have to look for is the one that does develop. That's the one that counts because it affects flying weather most, and often we can perceive its position before it winds up on a chart.

This flight illustrates how important the total cross-section of weather is to the pilot. The night before, all station reports were good but en route flying conditions turned out to be rather poor. The next morning, the station reports were bad but en route flying conditions were good. That strongly suggests that the low center was continuously one jump ahead of the forecasters.

JANUARY—KERRVILLE TO NASHVILLE

The Wicked Witch of Winter

The briefing for this flight was one that could lead to entrapment. The man in Texas said there was nothing on the map that might affect the route of flight, and in answer to a query about a rumored snowstorm to the west, he said that the radar

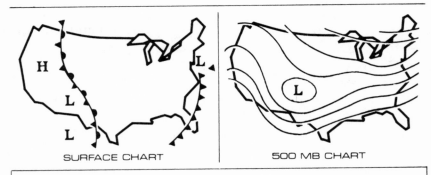

SURFACE CHART | 500 MB CHART

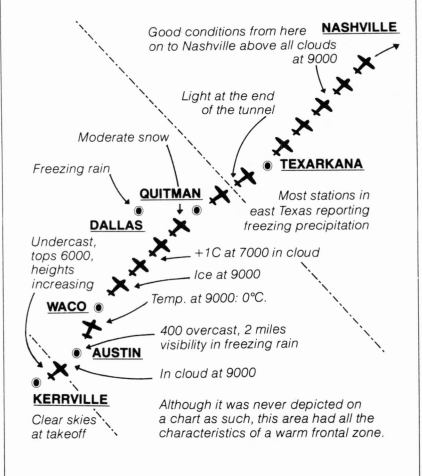

Good conditions from here on to Nashville above all clouds at 9000 **NASHVILLE**

Light at the end of the tunnel

Moderate snow

Freezing rain

QUITMAN

DALLAS

TEXARKANA

Most stations in east Texas reporting freezing precipitation

Undercast, tops 6000, heights increasing

+1C at 7000 in cloud

Ice at 9000

WACO

Temp. at 9000: 0°C.

400 overcast, 2 miles visibility in freezing rain

AUSTIN

In cloud at 9000

KERRVILLE

Clear skies at takeoff

Although it was never depicted on a chart as such, this area had all the characteristics of a warm frontal zone.

didn't show much of anything and he guessed that it wasn't materializing. San Antonio was 3,400 overcast and ten, Austin 4,100 overcast and seven, Dallas 10,000 broken and ten, Little Rock 1,500 scattered, 6,000 broken, and seven. He didn't have the Memphis and Nashville weather so I just filed for Little Rock and got a Memphis forecast to use for an alternate. The winds were light and southwesterly so I reasoned that I might not have adequate reserves for the 777 nautical miles to Nashville. I would stop at Little Rock in that case.

The skies were clear in Kerrville, and I was as optimistic as the briefer about a nice easy flight to the northeast.

Some clouds with tops about 6,000 showed up below my 7,000 foot cruising level shortly after takeoff. The air was choppy at 7,000, and the cloud tops moved on up to envelop that altitude as I flew along. A request for 9,000 was granted, and while that altitude topped all clouds, it didn't look as if it would do so for long. There was a gray line on the horizon. The air was rough at 9,000 feet, and the Stormscope in the airplane was indicating strong activity to the south. In checking weather I found that Austin had gone from 4,100 overcast down to 400 overcast with freezing rain. That man I talked with before takeoff didn't say anything about freezing anything, and where previously it hadn't been much of a consideration, icing became a big thing in my mind.

The temperature at 7,000 had been barely above freezing; at 9,000 it was +2 C. As the airplane moved into cloud at 9,000, the temperature dropped to about 0 C. Then, northeast of Waco, it dropped further and I started collecting some ice. By this time, Dallas had developed a case of freezing rain and there was much discussion of icing and turbulence on the frequency. The controller said that he had reports of ice below 6,000 and above 9,000 so I asked for 7,000 as a cruising level, to see if that would help.

Help it did. The ice slid neatly off the airplane at 8,700 feet, and at 7,000 it was about +1 C.

The temperature slowly dropped as I progressed northeastward. There was a lot of snow to fly through, and stations that weren't reporting freezing rain were reporting snow. The snow was almost heavy at times, but it would only make a thin white line along the leading edge and then dissipate when the airplane moved into lighter snow.

It was interesting to mull over the options at this point. The weather ahead was still okay, at least for IFR. Texarkana had the best, with 10,000 overcast; everything to the east of there was reporting from 1,000 to 1,500 overcast, some with very light snow. The stations in east Texas were all reporting freezing or frozen precipitation. What to do if ice started appearing at 7,000? It would probably be the last non-icy altitude, so changing altitudes wouldn't help much. The option of going somewhere else and landing would be open, and could carry with it a nice visit in Texas. It was apparent from the weather and the chatter on the frequency that something big was getting organized. (How big I didn't know until a few days later, when the low became one of the blizzards of '78 and dumped thirteen inches of snow on New York.) The best course of action seemed to be to continue. There were a lot of airports around, and if ice started to form, I could be on the ground rather quickly.

In reply to my questions about precipitation, the controllers kept saying that some seemed to be forming but that I'd be through it in another ten or twenty miles. That went on for 120 miles. I felt like the greyhound chasing a mechanical rabbit at the dog track. There were no pilot reports for me because I had been the only low altitude flight active in the area for the last three hours—all this on a day when good conditions had been indicated.

Close to Texarkana there was light at the end of the tunnel. And it turned out not to be a locomotive. I flew out beneath a higher overcast and could see a bright line ahead. It was clearing. What I had flown through was just like a warm front, probably in formation as a low pressure center to the southwest formed and developed a circulation. If it had been forecast to develop at that time and in that manner, I surely didn't know it from anything said by the FSS briefer. In fact, based on his information, I almost thought about flying VFR, to avoid circuitous IFR routing. That would have been a disaster.

After Texarkana, I had about fifty miles of clear skies and then flew up over an undercast. The temperature aloft was warm by this time, +4 C, but on the surface it was well below freezing. The absolutely flat-topped undercast persisted all the way to Nashville, and then on to the eastern side of the Appalachian mountains as I flew on after a fuel stop at Nashville. This was caused by an inversion that trapped the cold air near the surface and was an important illustration of how the vertical distribution of temperature can affect the flying weather.

One final point of interest about both this storm system and the system that harassed the two previous flights: The southwesterly circulation ahead of the storm system was not strong in either case. That might be construed as an indication of weakness on the part of the system, as the textbook picture of a deepening low is of strong and increasing circulation. These systems were just beginning as I met them, however, so the strength wasn't there even though the weather was rapidly becoming significant. That is something to watch for when you are in the breeding ground for surface low pressure centers. They can cause trouble as they start to form, and they can build into something ferocious later on. In this case things were helped along by the presence of a closed low aloft over northeast Texas on the morning of the flight. It helped provide the necessary upper level support to get things going.

FEBRUARY—
TRENTON TO POMPANO BEACH

Even the Weak Ones Count

A weak low center was off the coast, with low ceilings and visibilities inland. The forecast called for improving conditions by midmorning. No precipitation was indicated except in south Florida. Winds aloft were forecast to be light and westerly.

The flight was conducted at 6,000 feet, and was on top at the beginning. The actual cloud tops were about 3,500 feet but a haze layer extended to 6,000 feet. Despite the fact that marginal VFR conditions were forecast and reported, it was plain to see that VFR flying would have been all but impossible. There were occasional breaks in the clouds below, and lower scud and really cruddy visibility could be seen through these breaks. Norfolk had ground fog as we passed there.

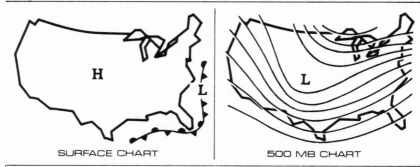

SURFACE CHART | 500 MB CHART

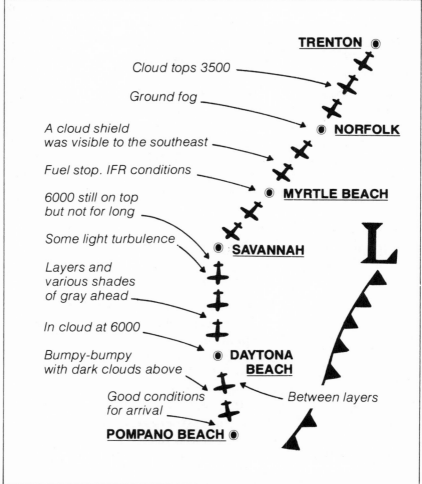

TRENTON

Cloud tops 3500

Ground fog

A cloud shield
was visible to the southeast

NORFOLK

Fuel stop. IFR conditions

6000 still on top
but not for long

MYRTLE BEACH

Some light turbulence

SAVANNAH

Layers and
various shades
of gray ahead

In cloud at 6000

Bumpy-bumpy
with dark clouds above

DAYTONA
BEACH

Good conditions
for arrival

Between layers

POMPANO BEACH

The cloud shield connected with the low center was visible to the southeast, and it was interesting to try and estimate our distance to the high clouds. As usual, they were farther away than they seemed and, in fact, we didn't actually fly under the higher clouds until long after a fuel stop at Myrtle Beach.

The air was quite stable this morning. The surface temperature was $+5$ C and the temperature at 6,000 feet was $+3$ C. There was probably a slight inversion at the cloud top level, though I didn't think to check for it on the climb or letdown. This inversion coupled with the light northeasterly surface flow around that low center to provide moisture for fog formation. It also provided the mechanism for holding that fog longer than forecast. Conditions had been forecast to improve by 10 A.M. but when we stopped at Myrtle Beach at 11, it was still below VFR.

Checking for the continuation of the flight, word was that the low was still in the same place and that conditions were quite low along the coast. Additionally, there was a Sigmet calling for embedded thunderstorms from Daytona Beach to Miami. The radar summary chart showed scattered thunderstorms and rainshowers from Daytona south. Winds were light southwesterly. The briefer said that there was no front shown on his map, but the daily weather map later published by the National Oceanic and Atmospheric Administration depicted a cold front just to the south of Key West.

Six thousand continued to top the clouds in the Charleston and Savannah area but the tops were rising. The cloud shield that I had been looking at all day was really off the coast of Florida; I flew under it at Jacksonville. There was some light turbulence there which increased as we progressed southward.

Just north of Daytona Beach the situation ahead became one of layers and various shades of gray. This was the area covered by the embedded thunderstorm Sigmet, but the radar controller said that he wasn't depicting any activity in the area. He did say that there had been some activity earlier but that it had dissipated and that the only weather return was now off the coast. At Daytona we were in cloud at 6,000 feet.

Some light chop developed in the vicinity of Melbourne and persisted as we flew out of cloud and into a between-layers situation in the vicinity of Vero Beach. It was like having flown through a little front and, indeed, had I been drawing a weather map, I'd have drawn a frontal line between Daytona and Melbourne. There should be something on the map when you fly from weather conditions of one sort through bumpy clouds and then into weather conditions of another sort.

The clouds ahead looked quite dark as we passed Vero and the higher clouds had the appearance of mammatocumulus. There were some heavy rainshowers falling from that higher layer, bases probably around 8,000 to 10,000, and there was some moderate turbulence in the vicinity of Palm Beach. The turbulence was related to wind shear as there was little up and down action. It was the sort that you often find in a weak frontal zone, reinforcing the feeling that there was a front in the area.

Conditions at Pompano Beach were good for a VFR arrival.

This day was a good example of how a rather benign weather system without any upper level support can create below VFR conditions over a wide area, and can harass the IFR pilot with the suggestion of embedded thunderstorms.

FEBRUARY—POMPANO BEACH TO MYRTLE BEACH

Florida Snow

My briefer this morning said that there were no features shown on the weather map. All stations were reporting VFR and winds were light and westerly. It was one of those light-weight weather briefings that gives an incomplete feeling, and the below average surface temperatures and the general appearance of the situation (cloudy) led to another call from the airport. Things had indeed deteriorated. Vero Beach was reporting 5,000 overcast and one mile visibility in rain and there was now a Sigmet calling for icing over the northern two-thirds of Florida. Forecast temperatures aloft over Jacksonville showed $+1$ C at 6,000 and -3 C at 9,000. I had filed for 9,000. This second FSS man said that from his experience, tops would be at 7,000. Surface temperatures were in the $+5$ C to $+10$ C range so we could always descend to melt ice. The final and best part of my briefing this morning came from friend John Becker, fixed-base operator at Pompano. He said that the

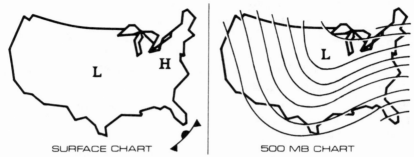

SURFACE CHART | 500 MB CHART

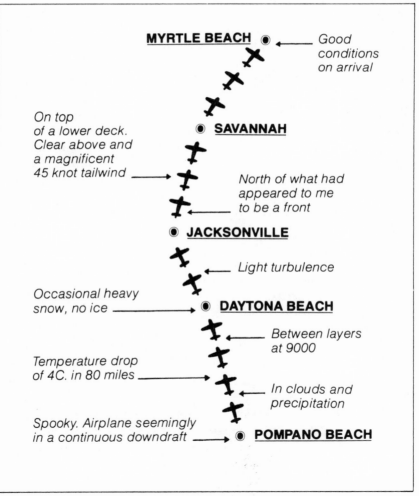

MYRTLE BEACH ⊙ ⟵ *Good conditions on arrival*

On top of a lower deck. Clear above and a magnificent 45 knot tailwind ⟶

⊙ **SAVANNAH**

North of what had appeared to me to be a front ⟵

⊙ **JACKSONVILLE**

Light turbulence ⟵

Occasional heavy snow, no ice ⟶ ⊙ **DAYTONA BEACH**

Between layers at 9000 ⟵

Temperature drop of 4C. in 80 miles ⟶

In clouds and precipitation ⟵

Spooky. Airplane seemingly in a continuous downdraft ⟶ ⊙ **POMPANO BEACH**

"Today" map had a front across northern Florida, with rain south of the front and good weather and high pressure to the north of the front. Why that front escaped the FSS man's map, I know not.

It was one of those spooky times for flying. There was a light jiggle in the clouds, and the airplane was not climbing well, as if there were subsidence in the area. After leveling at 9,000 feet, the cruising airspeed was off 10 knots as a further indication of some settling in the area.

The temperature at 9,000 started at +5 C and dropped quickly to +1 C. It took only eighty miles for that drop to occur. We were in cloud, and I felt that we might have an ice problem soon. The clouds and dropping temperature certainly argued in favor of ice. The four degree temperature drop in such a short distance argued against it by offering hope that the general situation would change. A rapid temperature change suggests entry into an air mass of different character.

The weather did change. At Vero we were between layers at 9,000 and then we were in and out of some very thin clouds. The temperature was −1 C and a trace of ice developed along the leading edges, but no more. North of Melbourne we flew into first light and then heavy snow but were clear of cloud so no ice developed. Snow in Florida is very unpleasant, even at 9,000 feet.

There was some light turbulence for a while at 9,000, and north of Jacksonville we were clearly north of the front. There was a lower cloud deck—Charleston was reporting 700 scattered, 1,200 overcast, and a half mile visibility, but Myrtle Beach was in good shape with 2,000 broken and five miles visibility.

Passing Charleston I checked the forecast temperature at 9,000 against the actual. The forecast called for −8 C and the thermometer read −1 C. The wind was a good deal stronger than forecast, too. I should have had an 11 knot tailwind; actually it was 45 knots. If it had been a situation where we were flying toward the area of weather instead of away from it, the large error in the temperature and wind forecast would have raised warning flags. Something was developing that they didn't anticipate. Looking at the next day's weather map showed that trouble did indeed develop. A low formed just south and east of a low on the 500 mb chart and moved eastward, spreading snow and cold temperatures through the southeast.

Back to this day's weather: It was a good example of what's often called the Confederate front. There almost has to be some weather encounter between cold country and warm country in the wintertime. If there isn't, chances are the weather in the warm place will be cooler than normal or the weather in the cold place will be warmer than normal. This day the activity was just farther south than usual. The Confederate front's more normal winter location is across southern Georgia, and its effect on VFR pilots has no doubt led to good occupancy rates in motels near the airports in the frontal zone. The "front" doesn't always show up on the weather charts, probably because it is often rather weak.

FEBRUARY—TRENTON TO LEXINGTON

North of a Weak Low

This flight was to be in the area north of a low in central Tennessee. The low was classified as a "weak" one, and was forecast to put an additional but gentle coating of snow across the midsection of the country. The reported and forecast weather was good as far as the mountains, where snow began. From there on west through Kentucky, the forecasts were for good ceilings and visibilities except for occasional 800 overcast and two situations in snow. I was going on farther west, to Dallas, and it would have been nice to make Bowling Green, Kentucky, which is halfway. I'd get there without much fuel, though, so the flight plan called for a stop in Lexington, Kentucky. Bowling Green was used as an alternate because its forecast was suitable for that. If speed and weather were both better than forecast, I'd go on to Bowling Green.

Flying across the north side of a low suggests tailwind but the forecasts were for westerlies at about 15 knots at 6,000 feet, the minimum en route altitude across the mountains. The low was apparently very shallow, right at the surface, where winds were out of the northeast. A low level VFR flight might have been faster, but the reported conditions didn't augur well for VFR.

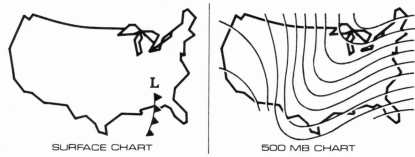

SURFACE CHART   500 MB CHART

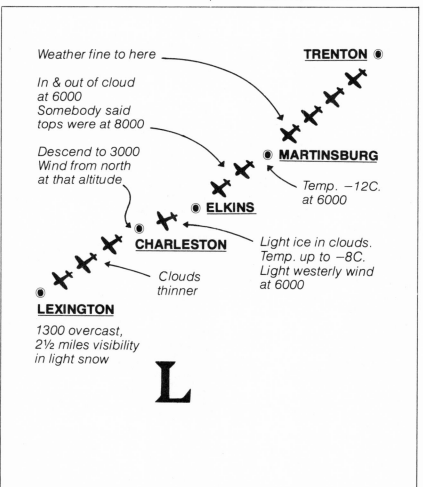

Weather fine to here

In & out of cloud
at 6000
Somebody said
tops were at 8000

Descend to 3000
Wind from north
at that altitude

TRENTON ◉

◉ **MARTINSBURG**

Temp. −12C.
at 6000

◉ **ELKINS**

CHARLESTON

Light ice in clouds.
Temp. up to −8C.
Light westerly wind
at 6000

Clouds
thinner

◉
LEXINGTON

1300 overcast,
2½ miles visibility
in light snow

L

The Appalachian mountains can be very inhospitable in murky weather.

Skies were clear as far as Martinsburg, West Virginia, and the temperature at 6,000 feet was − 12 C. I liked that cold temperature because it suggested freedom from icing conditions. The system was a slow mover so the clouds would likely be "old" stratus, meaning that they had been around long enough for the supercooled water droplets to become ice crystals in the cold temperatures.

At Kessel, I was in and out of cloud at 6,000 feet. There was a basic higher overcast, and some light snow. Closer to Elkins, the clouds at 6,000 became solid but there was still no icing. Tops were reported at 8,000 and I started to climb up there, but the wind was probably lighter at six. And why go up if there was no compelling reason (ice) to do so? I've always rather enjoyed flying in snow, anyway. It gives the airplane a nice cozy feeling.

Nearing Charleston, I started picking up some rather light ice. I was flying in cloud, and the temperature had risen to − 8 C, explaining the switch from ice-free to icy flight. There were still 130 miles left to fly, and the choice had to be made between going higher or lower. In favor of higher were the tops. In favor of lower was the fact that the wind there would be more favorable. There was, however, no haven at the surface in above-freezing temperatures.

A check of the weather showed Huntington to have 6,000 broken, 10,000 overcast; Lexington reported 1,000 broken, 7,500 overcast, 2½ miles in light snow. I opted for lower on the theory that I'd be below the cloud bases there, and just in some light snow that wouldn't add to the ice on the airplane.

The theory almost didn't work. In fact, I had to move all the way down to 3,000 feet to get out of the clouds and the ice. The wind at 3,000 was northerly, no headwind component, which made it quite an attractive altitude.

This flight was a good test of the theory of ice distribution around a low. When directly north of the low center we did accumulate a little ice. As we passed on to the northwest of the low, we got out of the ice. The simple reason for this was the thinning of the clouds. In fact, there weren't many identifiable clouds at Lexington even though they were reporting IFR conditions. From 3,000 the sun was dimly visible on an almost continuous basis, meaning that I was flying more in a collection of snow than in a collection of clouds.

I had the speed and fuel to go on to Bowling Green, but not with much reserve. And a weather check before decision time showed them reporting 800 overcast with three-fourths of a mile visibility in light snow, blowing snow, and fog. Lexington was reporting 1,300 overcast with 2½ miles in snow. As mentioned, though, the condition was more one of obscurement than overcast. I landed at Lexington for fuel.

This was a rather weak surface low but it still managed to deposit nine inches of snow on parts of Tennessee. Its effects were widespread, too. The snow started at Kessel and lasted to Bowling Green, a distance of over 350 nautical miles. The low finally had some effect on the upper winds, too. Where the wind aloft was light westerly at 6,000 north of the low center, it was pulled around to northerly on the backside of the low. This did nice things to our groundspeed for quite some time after the fuel stop at Lexington but, alas, the wind shifted back to westerly in Arkansas, slowed us down, and necessitated another fuel stop before we reached Dallas.

In retrospect, the decision to descend instead of climb when the light icing started around Charleston could be flawed. Climbing would have been a sure cure, descending was a possible cure based on the ceiling being as reported. And it almost didn't work. The lure of less wind component at a lower altitude was strong, but it really shouldn't have influenced the decision. It would have been better to go higher.

This low, though weak, did have a lot of support aloft. There was a strong trough at the 500 mb level, moving eastward very rapidly and steering the low toward the east. Perhaps the upper air pattern explains why there was more snow than expected in Tennessee.

MARCH—TRENTON TO TRENTON

Show Me the Way to Go Home

The early morning briefing for a flight to Dayton positioned a low pressure center in Tennessee with a warm front to the northeast and a cold front trailing to the southwest. Addition-

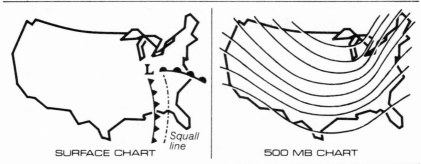

SURFACE CHART   ·   500 MB CHART

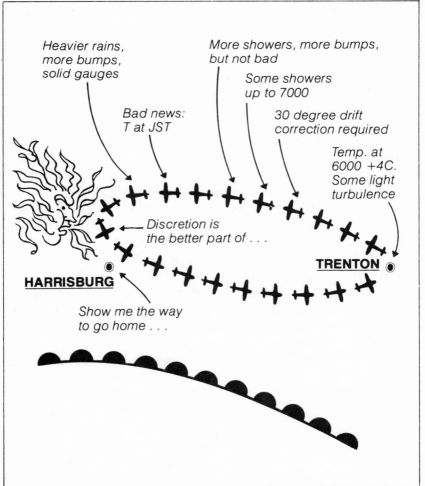

Heavier rains, more bumps, solid gauges

Bad news: T at JST

More showers, more bumps, but not bad

Some showers up to 7000

30 degree drift correction required

Temp. at 6000 +4C. Some light turbulence

Discretion is the better part of . . .

Show me the way to go home . . .

HARRISBURG

TRENTON

ally, an east-west stationary front was to the north, along the Canadian border. Forecasts called for broken clouds at about 1,000 feet and an overcast at 2,500 feet as far west as Pittsburgh. Over in Ohio, the forecast norm was 500 feet, with occasional 200 obscured in fog. Indiana was a little better: Indianapolis called for 800 overcast and three miles. In retrospect, the problems of this day were caused at least in part by the fact that the briefer gave a poor description of the weather map. Contrast his word picture with the surface chart as it actually existed at 7 A.M.

There were numerous Sigmets for turbulence—moderate below 12,000 feet with occasionally severe turbulence below 18,000 feet over southern Ohio. What was different about southern Ohio? No reason was given, and I could only imagine that this turbulence was expected in connection with the approach of the low center.

The freezing level was reported at 6,000 over Ohio, sloping up to 12,000 over Virginia.

The radar summary chart showed nothing, according to the briefer, and no thunderstorms were forecast. Very strong winds aloft were forecast, southerly to south-southwesterly at 45 knots at 6,000 feet, stronger and slightly more westerly aloft.

I asked the briefer to describe the 500 mb chart but it was apparent that he didn't even have a rudimentary knowledge of 500 mb charts. He was unable to understand that I wanted to know the location of the southern tip of the trough and when I asked him about the wind he gave it to me for some altitude in the thirties. I finally got the forecast for 18,000 foot wind at Pittsburgh—southwesterly at 64 knots. Thus, the low would probably track to the northeast. From central Tennessee, that would send it through central Pennsylvania.

My destination was Dayton with Indianapolis as an alternate. I felt it reasonable to have a look; it was all very legal from a forecast and a fuel standpoint. But I had some doubt about flying the flight as planned because of the possible collision course with the low center. I was using the apparent presence of an inversion (based on actual surface temperatures and forecast temperatures aloft) as an indication that the turbulence across the mountains wouldn't be so bad. Inversions do tend to reduce turbulence.

There was a strong inversion, too—from +3 C on the ground up to +10 C at 2,500 feet. But it wasn't a real inversion. It was to be expected in climbing from the snow-covered frozen ground into moist air moving from the south. From 2,500 to 6,000 the temperature dropped back to +4 C, not a sign of instability but not one of stability, either. Of course it would be foolish to expect much stability in the area to the northeast of a low center.

There was occasional light turbulence at 6,000. In some areas the tops were at about that level but these tops were of cumulus clouds and were very irregular. The cumulus seemed to build occasionally into the higher overcast.

The feeling was one of being in a very bothered sky. The wind was strong, too. A 30 degree drift correction to the left was required to track the airway and the groundspeed was 20 knots below the true airspeed. That was to be expected from the forecast winds. The temperature at 6,000 was a bit above forecast.

The controller had to move me up to 7,000 feet to clear some other traffic and the temperature there was +1 C, or three degrees lower than at 6,000. Was that some localized thing, or was it a good indication of greater instability?

There were a lot of other airplanes out flying and there was a good bit of discussion about the weather. A commuter airline pilot reported moderate ice while flying at 12,000 feet and requested a lower altitude with some urgency in his voice. Several jetliner pilots reported that they were descending at reduced speed because of turbulence, and a Cheyenne pilot complained of continuous turbulence at 7,000 feet.

My situation seemed good by comparison, at least for the moment. The groundspeed wasn't too bad, the temperature was above freezing, and the bumps weren't overly bothersome.

There was a big question about what lay ahead, though. I hadn't yet reached the area where I'd feel the effects of the strong wind flow over the mountains, and that could extend up much higher than 7,000 feet, which I considered the lid because of icing above that level. Still, I was game to keep on, to take at least a small bite of the apple.

The Center controller had no low level pilot reports, but he said he'd be switching me over to Harrisburg Approach Control in a minute and they might have something.

Harrisburg had nothing. And in answer to a query about weather return on his scope, the controller said that his radar was set to blank out all weather.

I got permission to switch over to the FSS to test their information and found it quite interesting. Johnstown, toward which I was flying, reported zero-zero with a thunderstorm in progress, cloud to ground lightning in all quadrants, wind southerly at 20 knots. Other airports in the area were reporting 100 feet or zero-zero in fog. The FSS had no pilot reports.

I questioned the veracity of a report that said the visibility was zero yet lightning could be seen in all quadrants; this definitely called for further investigation. Back at Approach Control, the man agreed to change the adjustment of his radar and have a look. Sure enough, in a minute he said that there was a line of weather to the west of Harrisburg that extended off his scope.

My flight conditions were deteriorating somewhat, with heavier rain and more turbulence in the showers. Experience, and a basic understanding of meteorology, suggested that a wet and bumpy condition would worsen while flying toward a low.

At this point I had to conclude that all the information gathered before takeoff was obsolete. No thunderstorms had been forecast, yet one was reported at Johnstown and there was weather return on radar. Perhaps the warm front was active with very unstable warm air overrunning and spawning the activity. Johnstown had to be near or north of the surface position of the warm front to maintain zero-zero in the storm (warmer rain falling into cooler air right at the surface) and I suspected that the frontal surface would not be very far above the ground in that area. Also, I was probably rather close to the low center, which would have moved northeast from Tennessee. Close to the low and in its path—not a good place to be.

The thunderstorms would be embedded in other clouds. At least I was on solid instruments and what I had seen so far suggested that between-layers flight would be impossible.

If the country had been flat, the temptation would have been to continue at the minimum en route altitude. But the country wasn't flat and the MEA wasn't low. And, the lower I flew, the more effect from the strong wind over rough terrain.

I felt boxed in from above by ice, from below by terrain and turbulence, and from ahead by a line of thunderstorms. Those storms were probably not too strong, but in a light airplane a storm is a storm is a storm.

The decision to retreat is never easy but if I had continued I would have soon been over an area where the airports were reporting conditions below IFR minimums. Options would be diminishing in an area of worsening weather. Thunderstorms were reported where none were forecast. All that was a bit much so I turned around and went home, to be a couple of days late.

There was a pretty good trough aloft this day, with a jet stream on the southeast side of the trough and just to the southwest of Pennsylvania. The low was quite deep and, as mentioned in the beginning, it was quite far from where the FSS briefer had positioned it. I had an inaccurate picture before takeoff, and once en route the only thing certain was that things were getting progressively worse.

MARCH—TRENTON TO DAYTON

Snowy Day

The weatherman mentioned a winter storm watch on TV the night before, and my early-morning briefer said that a low center was over central Tennessee, moving very slowly east-northeastward. It was expected to move off and then up the coast, as a late season northeaster. The weather along the coast was good but there was a wide area of snow across the Appalachian mountains. Phillipsburg, Altoona, Johnstown, and Pittsburgh were all reporting snow, with some conditions below IFR minimums. The weather improved west of Pittsburgh. The winds aloft were forecast light southwesterly at 6,000, increasing some in velocity with altitude but still remaining moderate. The forecast temperature was -8 C at 6,000 and -13 C at 9,000. There was one Sigmet current. It called for moderate mixed icing in clouds and in precipitation over Ohio. The briefer said

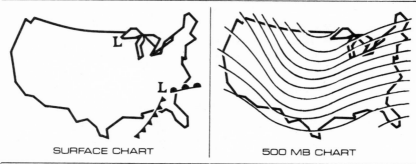

SURFACE CHART 500 MB CHART

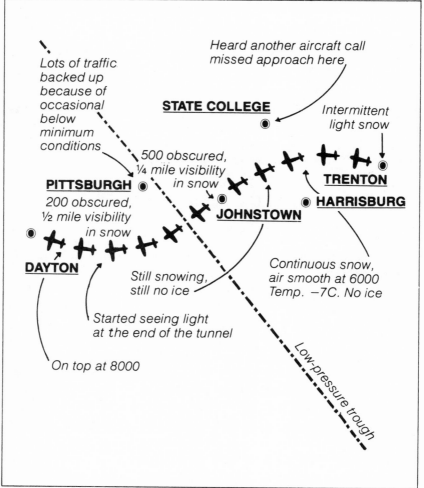

Lots of traffic backed up because of occasional below minimum conditions

Heard another aircraft call missed approach here

STATE COLLEGE

Intermittent light snow

500 obscured, ¼ mile visibility in snow

PITTSBURGH

200 obscured, ½ mile visibility in snow

TRENTON

HARRISBURG

JOHNSTOWN

DAYTON

Still snowing, still no ice

Continuous snow, air smooth at 6000 Temp. −7C. No ice

Started seeing light at the end of the tunnel

On top at 8000

Low-pressure trough

the radar report didn't show much out that way so he assumed the tops would be low. While listening to the news on the radio, I heard a station mention a low pressure system in Ohio, which was added to the list of possibilities for this day.

If the forecast of light southwesterly winds was correct, the low was either not strong or not close. If the forecast of low temperatures aloft was correct, there shouldn't be much ice. I mentally questioned such cold temperatures with a southwest wind, but anything is possible. I should have tried to get some picture of the 500 mb chart but I was talking with the same briefer who had been unable to understand a request for this information a couple of days before—so I didn't even try.

Six thousand was the chosen altitude, and the snow was quick in coming. At first it was intermittent and light, but by the time I was eighty miles west of Trenton, snow was continuous. I didn't have the feeling that I was in cloud; it was just snow that obliterated everything. It was like flying inside a big (and full) milk bottle.

I heard an aircraft call with a missed approach at University Park airport, just north of my route. That prompted a slightly overdue weather check. It was really snowing over the mountains. All stations were giving obscurement in snow, with a fourth to a half a mile visibility. The weather west of Pittsburgh was fine, though, and there was no ice at 6,000 and there were no pilot reports of ice.

I'd feel an occasional jiggle, and could note the slightest frosting of the leading edge after passing through this ever so light turbulence. It must have come in the tops of scattered clouds that formed in the snow.

The groundspeed was 135 knots, very close to the true airspeed, and I couldn't perceive much drift correction as I tracked the airway. The wind was either virtually calm at 6,000 or perhaps light southeasterly, which is quite unusual for mid-March—especially during a snowstorm.

When I heard a pilot report of solid IFR conditions from the surface to 24,000 feet, I thought back to the briefer's idea that the tops over the mountains would be low. Sure didn't happen that way.

The temperature at 6,000 was close to forecast, at −7 C.

And as I progressed, there was but one hint that the good flying conditions wouldn't continue. I checked weather with Morgantown Radio, to find it much the same as the previous hour, but they added a pilot report of icing northwest of Morgantown. I'd be passing northwest of Morgantown so I asked if that was a report or a forecast. He assured me that it was a report. After I switched back to the traffic controller, I quizzed him thoroughly to find that they had had no reports at all of icing anywhere in the area. I took that as a better indication than the FSS report because the Center controller is usually the first to know of any icing problem. Pilots just aren't bashful about discussing ice with controllers.

I was detoured around to the south of Pittsburgh, because of aircraft holding for landing at Greater Pitt. Conditions there were varying from below to above minimums and this was apparently causing quite a tie-up. I was glad I didn't need to stop there.

In adjusting the altimeter to each new setting, Johnstown was the place where the setting started moving higher rather than lower. Even though surface conditions beneath remained very poor, the sky seemed to brighten. It had the appearance that is typical of the backside of a low. I couldn't see whether or not the sun was dimly visible because it would have been hidden by the high wing, but I imagined that it would be.

I started seeing some ground southwest of Pittsburgh and the high overcast ended shortly afterward. There was a stratocumulus deck to the west, with tops about 6,500 and I had to climb on up to 8,000 feet to avoid ice.

The weather this day was influenced by very strong winds at the 18,000 foot level. The primary low was about halfway up the east side of the trough on the 500 mb chart and was moving rapidly northeastward. There was another low, a very weak one, shown in Wisconsin and the area of snow was along a trough that ran between the two lows. The trough wasn't depicted on a chart but there's usually one there between a pair of lows.

MARCH—DAYTON
TO OLATHE (JOHNSON COUNTY)

Snow Burst

The briefing for this, a continuation of the previous flight, suggested a totally routine ride to Olathe, Kansas. "There's nothing out there," said the briefer. Even the winds aloft sounded pretty good—northwesterly at 15 knots.

And indeed there was nothing but blue sky and some lower clouds below as I flew across Indiana and Illinois at 6,000 feet. But Missouri was something else. At St. Louis some cloud tops were beginning to poke above 6,000, and clouds to the west started assuming a dark and pregnant look. Things got even more interesting to the west of St. Louis, and when I quizzed the FSS at Columbia about weather on to the Kansas City area, they said it was pretty good, but not as good as forecast. There were some snow showers around that were lowering ceilings and visibilities. I was busting through some rather bumpy clouds at 6,000 and was getting a trace of ice in each one. The temptation to cancel IFR and go VFR down lower was rather strong, but the thought of a lot of snow showers in the Kansas City area was enough to keep me hooked onto the IFR system.

The wind was stronger than forecast—it was doing about 30 knots worth of harm to the groundspeed—and the only mental picture I could put together was one of instability behind a cold front causing the showers.

The closer I got to Kansas City, the more pleased I was with an IFR clearance. Aircraft arriving VFR were seeking IFR clearances, and my destination, which had been forecasting 4,500 broken and 10,000 broken with unlimited visibility, was now advertising 1,200 broken, 4,000 overcast, and two miles visibility with ice pellets. (I never did like the word "pellets" in a weather report because I used to have an air gun which shot pellets that could make holes in something like an airplane.)

The controller approved some deviations as I tried to aim for the lightest spots in the snow showers, and he approved a descent to 4,000. I needed to stay out of increasingly frequent shots of ice at 6,000 while flying through the snow shower clouds. Tops in the area were reported at 12,000 to 15,000

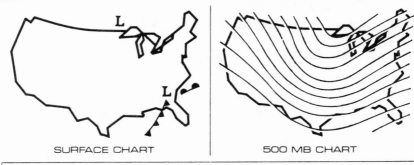

SURFACE CHART | 500 MB CHART

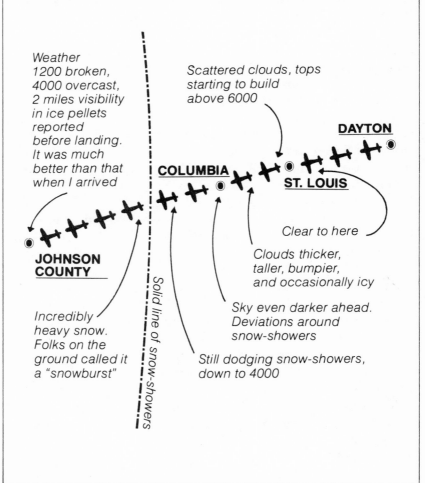

Weather
1200 broken,
4000 overcast,
2 miles visibility
in ice pellets
reported
before landing.
It was much
better than that
when I arrived

Scattered clouds, tops
starting to build
above 6000

DAYTON

COLUMBIA

ST. LOUIS

**JOHNSON
COUNTY**

Solid line of snow-showers

Incredibly
heavy snow.
Folks on the
ground called it
a "snowburst"

Clear to here

Clouds thicker,
taller, bumpier,
and occasionally icy

Sky even darker ahead.
Deviations around
snow-showers

Still dodging snow-showers,
down to 4000

feet, with some higher. At times my ride was uncomfortable; the ride was better at 4,000 which was below the bases of most of the clouds.

Moving into the Kansas City area, Approach Control issued a vector heading which I dutifully followed. The sky ahead assumed a rather dark appearance but there was no visible best way to detour and I continued on the issued heading.

The turbulence increased a little as I flew into the dark area and then incredibly heavy snow commenced. It was hitting the windshield with a "whumph" sound and the noise of the snow hitting the airplane was actually louder than the engine sound. There was a fairly strong updraft at the beginning, too.

Some frozen stuff was collecting at the leading edge, but it was more impacted snow than ice. Still, I thought I'd try for a lower altitude, 3,000 feet, which was approved. The snow continued for another couple of minutes and then I flew out on the backside of the activity. Looking back, it appeared to be a line, and after I landed some folks referred to what had moved through the area as a "snow burst." I hadn't heard that term before but it was very descriptive.

Two things on the weather chart offer some explanation of the outburst of snow in Missouri. The trough at the 500 mb level was right over the area, offering some upper level support, and the temperature aloft was very cold. It was -35 C at 18,000 feet compared with $+4$ C on the surface, which is a good indication of instability.

MARCH—
BOWLING GREEN TO LITTLE ROCK

Classic Frontal Model

The first briefing of the day set the stage for a routine trip from the east coast over about half the distance to Little Rock, Arkansas. Bowling Green, the farthest it would be possible to go for a fuel stop, was reporting rather good conditions and was forecast to remain good with only the chance of below VFR conditions in rainshowers or thundershowers. The map described by the briefer had a low situated over Arkansas with a stationary front to the northeast and a cold front trailing to the

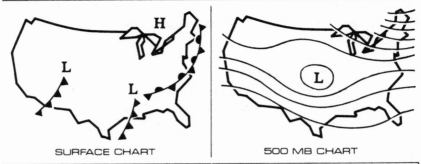

SURFACE CHART | 500 MB CHART

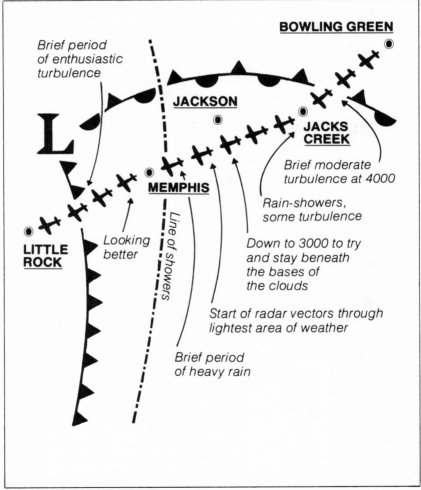

BOWLING GREEN

Brief period
of enthusiastic
turbulence

JACKSON

**JACKS
CREEK**

L

Brief moderate
turbulence at 4000

MEMPHIS

Rain-showers,
some turbulence

Looking
better

Down to 3000 to try
and stay beneath
the bases of
the clouds

**LITTLE
ROCK**

Line of showers

Start of radar vectors through
lightest area of weather

Brief period
of heavy rain

southwest. I'd be flying north of the stationary front all the way to Bowling Green and even the winds were forecast to cooperate that far, with light and variable on tap at 6,000 feet.

The flight to Bowling Green was as nice as the forecast, and the only note I made on the elements was about the temperature at 6,000 feet being 5 degrees C above the forecast. It also struck me that the winds were light southeasterly rather than light and variable, and perhaps that related to the higher than forecast temperature.

I went into the FSS at Bowling Green to get weather information for the continuation of the trip. When I told the FSS man that my flight would be to Little Rock, he greeted me with one of those bureaucratic "oh boy" exclamations. It was a difficult briefing because he wanted to read rather than let me examine the reports but I got what I thought I needed. The low was still in Arkansas, and it appeared that I would be flying through first a stationary front and then a cold front. Ceilings were good to the stationary front and then they were from 1,000 to 2,000 feet on to Little Rock. The radar chart showed a large area of rain over the route with tops to 18,000 feet. Jackson, Tennessee, was reporting a thunderstorm which didn't show on the radar chart. Memphis was forecasting the chance of a thunderstorm and Little Rock, behind the cold front, was forecasting 2,000 overcast, three miles visibility, occasional rainshowers. Winds were forecast for the southwest at 20 knots. In all, it didn't look like such a bad deal because the low lacked the strength of the springtime storms that spawn squall lines. I filed an IFR flight plan, to cruise at 4,000 feet. That altitude was selected to minimize the effect of the headwind as well as to try and stay in visual meteorological conditions as much as possible—to be able to eyeball any thundershowers that might develop.

The surface wind was fresh and from the northeast at Bowling Green. The temperature was +16 C; Nashville, but fifty miles to the south, had a strong southerly wind and +24 C. The first front was definitely there and I didn't have any illusions of a smooth ride. Indeed, there was some light to moderate turbulence about forty miles southwest of Bowling Green. I was still clear of cloud at 4,000, but the bases did seem to be dropping. In the vicinity of Jacks Creek, bases were down to an estimated 5,000 and there was continuous light turbulence at 4,000 feet.

Passing south of Jackson, Tennessee, 4,000 was right in the bases of the clouds and there were some rainshowers. The turbulence wasn't bad, though. Occasionally I could see a rip in the cloud cover, blue sky above, and this prompted me to inquire about tops. General cloud tops were reported at 22,-000 by an airline jet, so I didn't give much thought to climbing.

The controller was able to approve 3,000 for me when I was southwest of Jackson and this was again below the clouds. The view was typical of a springtime southwesterly flow in the south. Lots of rainshowers, rather angry looking clouds, and bumpy air beneath the clouds. The difference between this and a real springtime storm was the weak circulation around the low, which was both slow moving and weak.

The ceiling lowered below 3,000 and I was on instruments passing north of Memphis. The rain was more continuous and, when questioned, the controller said that a north-south line of weather appeared to be forming ahead. I asked for a vector that would take me through the lightest area, and this required only a 20 degree heading change.

Moving through the line involved some moderate to heavy rain for about three or four minutes, plus a little light turbulence. There just wasn't much vertical activity at 3,000 feet. I had the airplane slowed to maneuvering speed for penetration of the weather, and the airspeed stayed neatly put and there was only a slight tendency to climb.

Just past Memphis I flew into the quite different conditions behind the cold front. The ground was occasionally visible below, through scud and rain, and the view was a good illustration of a VFR trap. The weather reports were all showing tolerable VFR conditions—the worst reported was 1,100 overcast and three miles in rain, which isn't impossible in flat country—but there wouldn't have been any VFR through the developing line of weather. Some of that stuff was right on the ground. The development hadn't really been forecast, either. The briefer in Bowling Green didn't mention it, so it would probably have taken a pilot flying VFR by surprise. One more proof that weather is what you find, not what you expect to find.

The controller sent us back up to 4,000 where I was on top part of the time and between layers part of the time. The rest of the time the airplane was punching through bumpy cumulus behind the front and at one point the turbulence became rather enthusiastic for about a minute. That probably marked the penetration of the cold frontal slope. The surface wind at Memphis was southerly; Little Rock's wind was westerly at 20 knots. The weather there was good VFR for our arrival.

The line of weather at Memphis never developed into anything severe, but some thunderstorms did sprout. The "Today" program that morning had mentioned the chance of severe thunderstorms in Mississippi but those never materialized.

Looking at the 500 mb chart, you can see that there wasn't a lot of upper level support for the development of a strong low. The flow to the east of the low aloft was not too strong and there wasn't a lot of cold air being moved south of the low and back northward.

This low center continued to move very slowly. The day after this flight, the low was in southern Illinois, still rather weak. But it spawned an ice storm that closed Chicago's O'Hare airport for only the fourth time in history and there was at least one general aviation accident in the vicinity of the low center that might have been ice-related. Two days later, the low moved back down to eastern Tennessee, under that low aloft that was meandering lazily eastward. Then it got quite wet and dumped up to three inches of rain on the northeast before going out to sea.

7

Flights: Second Quarter
April, May, June

This is a lovely time of year, for flying or anything else. True, there are some big storm systems that spawn such meteorological delights as tornadoes and severe thunderstorms, but these are easily identified and avoided. It's hard to hide those tall and awesome storms, and the pilot who falls victim to one has almost always flown past a host of warning signals from the people on the ground and, more importantly, from the view out the window of the airplane.

Summer comes to the south early in this period, and to the north toward the end of the period. The systems tend to be more diffuse in the summertime, so a loss of clarity in the various factors creeps northward during this time. That's not to say there can't be some wild weather—one of the most memorable tornado outbreaks in recent times moved across the southeast on Memorial Day a few years back—but for the most part it's a time of partly cloudy with occasional thunderstorms in the south, especially in June.

Another weather factor comes into play as the weather warms. Our airplanes don't perform as well in the warmer temperatures as they did in the cold, and we have to start paying close attention to the surface temperature for takeoff when the field length is the least bit marginal. It's a good time of year for birds, too, so watch out for those nests in the cowling.

This time of year is excellent for aerial sightseeing because most weather systems are still moving reasonably well, often providing good visibility after they pass, and the only thing prettier than the countryside in the springtime is the countryside in the fall.

APRIL—SHELBYVILLE TO WICHITA

First Twister of the Year

The briefer said that there wasn't a low on his map but that there was a warm front across southwest Missouri and central Kansas. All stations were reporting good weather, but St. Louis was reporting thunderstorms northwest through northeast and there was a Sigmet and tornado watch out for the general area. According to the radar chart, most of the activity was to the north. The low level winds were from the south to southeast; up higher the winds were from the southwest. The circulation was moderate, with surface winds at Wichita southerly at 20 with gusts to 30. (Such values would be considered strong except in the Great Plains, where 30 knots isn't anything to write home about.)

The best deal appeared a low level flight. The winds would be favorable there, and general VFR conditions meant that it would be possible to eyeball any storms.

Southwestbound at 4,500 feet, I found relatively smooth air and lots of chatter on the radio about weather. All the significant stuff was north, up closer to Chicago, but it was interesting to listen to people as they sought information on possible storm-free routes. Nobody seemed to be having any trouble getting where they were going.

There was no weather of consequence until I was past central Missouri. Then the higher clouds took on an angry look. They almost qualified as mammatocumulus. A line of showers appeared ahead, but the rain was falling from higher clouds, no lower clouds were involved, and a pilot report from an airplane that had just passed through the area revealed that only a touch of moderate rain and some light turbulence prevailed in the line, at least at low altitude.

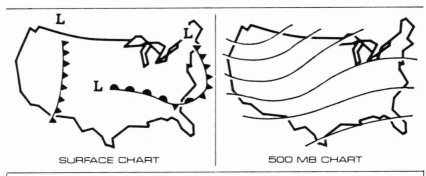

SURFACE CHART | 500 MB CHART

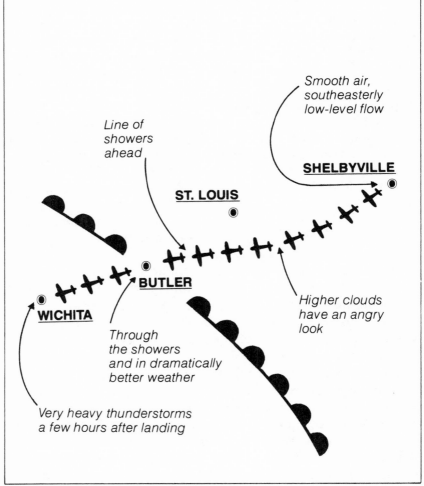

Smooth air,
southeasterly
low-level flow

Line of
showers
ahead

SHELBYVILLE

ST. LOUIS

BUTLER

Higher clouds
have an angry
look

WICHITA

Through
the showers
and in dramatically
better weather

Very heavy thunderstorms
a few hours after landing

After flying through the line of showers, the appearance of the situation changed markedly. On the east side, there was a lot of haze and the higher clouds were ill-defined except for those that had the mammatocumulus look to them. On the west side, the visibility was absolutely unlimited. The higher clouds were more sharply defined. I could see a break in the high clouds and I felt I could actually see a curve in them, where they moved from an easterly to a northeasterly direction.

I hadn't flown far beyond the line of showers when the air became quite turbulent at my 4,500 foot cruising level. The choice was between going higher and accepting more headwind, or going lower to minimize the wind and take the bumps. The Wichita weather had a bearing on the decision. The reported 2,300 overcast meant that a higher altitude would involve IFR. It would be quicker to stay low, and that was the original decision. The second time I strained against the belt in turbulence, though, the IFR decision was made.

In the notes made during the flight, I reasoned that the line of showers had to have been a warm front. The surface wind shifted from southeasterly to southerly and increased in velocity, the temperature was higher to the west of the line, and only the pressure defied the diagnosis. It was a bit lower west of the line than east. Still, in my mind I positioned a warm front back there and decided that I was now operating in the warm sector.

The landing at Wichita was uneventful. The tornado watch was still on, though, and while I felt as if my earlier sighting of blue sky to the west meant that any bad stuff would be southeast of Wichita, I accepted a friend's kind offer to use a little corner of his big hangar. I was later quite pleased that my airplane was in the hangar.

I might have been correct about the shower area defining some sort of warm front, but I was sure wrong about the thunderstorms missing Wichita. An hour or so after landing, I watched the news and saw, on the TV station's nifty color radar presentation, a very strong line of thunderstorms to the west. These moved on Wichita and passed through at about 7:15, complete with heavy rain, strong winds, and tornado sightings. There was no damage in Wichita, but the storm system later tore up an airport in eastern Kansas and did assorted damage elsewhere.

In looking at the actual weather maps for the day, it is hard to see the basis for issuance of a tornado watch early in the day, for the time that I was flying from Shelbyville to Wichita. There was nothing going on aloft or at the surface to suggest severe weather. I suppose there was the chance for some development there, but it looks pretty sparse and illustrates how we have to take such watches with a grain of salt. A tornado watch certainly suggests that it is a time to be alert, but I won't cancel a flight because of one as long as the weather is such that I can see, or I have weather avoidance gear in the airplane. (On the other hand, a tornado *warning* is an indication that one of the beasts has actually been sighted.)

What brought the Wichita storm was the rapid development of what had been a rather vapid low in western Kansas. All charged up, the low and associated cold front took off eastward, with upper level support provided by the trough shown on the 500 mb chart. It was a delayed action storm—the original tornado watch for the area expired long before the severe weather got there.

APRIL—TRENTON TO SAVANNAH

Springtime Warm Front

There was a low in Kansas with a warm front eastward, shown in the vicinity of Savannah at 1 A.M. All the forecasts along the route called for VFR conditions with the chance of rainshowers only at Wilmington, N. C. That was logical. Wilmington would be north of the warm front; if anything developed, it might well be there. The radar summary chart showed nothing and there were no Airmets or Sigmets. The wind aloft below 10,000 feet was forecast to be westerly, shifting to southwesterly in Georgia. I got all this information at 5:30 in the morning, though, which made it suspect. The weather system tends to sleep at night and to be slow awakening in the morning.

Aloft, the first weather item was a pleasant surprise. There was almost no headwind component. An unpleasant surprise came with an area of snow in southern New Jersey. It was falling from a layer based at about 8,000 feet (I was cruising at

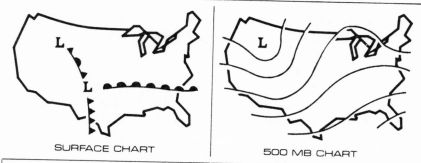

SURFACE CHART | 500 MB CHART

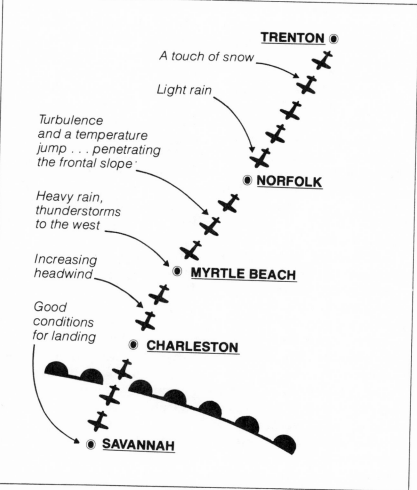

TRENTON ⊙

A touch of snow

Light rain

Turbulence
and a temperature
jump . . . penetrating
the frontal slope·

⊙ NORFOLK

Heavy rain,
thunderstorms
to the west

Increasing
headwind

⊙ MYRTLE BEACH

Good
conditions
for landing

⊙ CHARLESTON

⊙ SAVANNAH

6,000) and was a sure sign of instability in overrunning air associated with that warm front to the south.

At Norfolk there was some light rain, freezing rain really, as the temperature was −2 C. It didn't amount to much, though. A bit farther south there was some more light rain and then a period of turbulence. A substantial temperature increase was associated with the turbulence so I deduced that this marked the slope of the warm front and that the rest of my trip would be in the warm air.

Six thousand became a rather uncomfortable flight level. I was in cloud most of the time, the bumps meant the clouds were cumulus, and the development of cumulus above the frontal slope was a sure sign of instability. That is a strong suggestion to be prepared for what logically comes next—the development of showers or thunderstorms.

I tried 4,000 feet for smoother air but it didn't help much. The rain became heavy in the vicinity of Myrtle Beach and remained quite heavy for a number of miles. The raindrops were big ones, and I strongly suspected that a low pressure system was forming off the coast, at Cape Hatteras. This turned out not to be true, and if I had seen the 500 mb chart it would have been obvious that things were not ripe for low development out there. However, the big drops and the area of rain was much like I had seen in this area before when a low formed unexpectedly.

The groundspeed was still within 10 knots of the true airspeed at Myrtle Beach. The rain was heavy and pilots were talking of thunderstorms on the frequency. The Stormscope in the airplane indicated a developing storm not far off to my right, so there was indeed something to talk about.

The rain, and the thunderstorm activity that was developing, ended rather abruptly south of Myrtle Beach. Checking the weather ahead, I learned that Charleston was okay but with a strong east wind at the surface. Savannah was reporting excellent weather with the surface wind from the south. The surface position of the warm front was obviously between Charleston and Savannah. The groundspeed was 30 knots under the true airspeed when I passed Charleston, and the wind was strong southerly, as shown by drift to the right. On arrival, conditions at Savannah were beautiful and typical of the warm sector in this part of the country. They called it 2,000 broken with seven miles. The clouds were cumulus, like castles in the sky.

This wasn't a particularly notable warm front, but it amounted to a bit more than the forecasters expected. The thunderstorms continued to develop on that front all day, but there was nothing severe or well organized because there wasn't any upper level support for the activity.

Even though the slope was very shallow—the warm air at 6,000 feet was about 300 miles north of the identifiable surface position of the front—the weather associated with it was worthy of every consideration, and it affected a rather large area. It was a good lesson on how the frontal line isn't as important as the frontal *zone.* Even though the line on the map was far south, the weather started with that area of light snow in Jersey, escalated to a point south of Myrtle Beach, and then ended before I reached that hallowed line on the map, the surface position of the front.

APRIL—GULFPORT TO BRISTOL

Squall Line

This day got off to a bad start. The briefer prefaced his formal remarks with an exclamation, which to me is an indication that the person is more interested in dramatics than dispensing information. There was a lot to talk about, too. A solid line of thunderstorms was oriented from north of New Orleans, up over Hattiesburg and Montgomery, and as far to the northeast as the radar eye could see, according to the radar summary chart the man was using. On the weather radar at Mobile, the actual real time picture showed a line from north of Mobile to the northeast. The briefer tried to suggest that there were two lines of thunderstorms—the one on the radar summary chart and the one on his radar—but that didn't seem very likely. The summary chart was old, and the line had just moved to the east.

I was going home to New Jersey, and the normal fuel stop would have been Greensboro, North Carolina. The weather up that way was terrible, though, and was forecast to remain terrible with low ceilings and thunderstorms. The synopsis showed

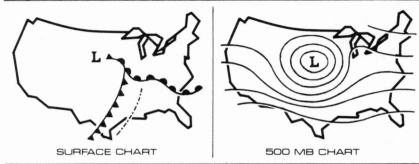

SURFACE CHART | 500 MB CHART

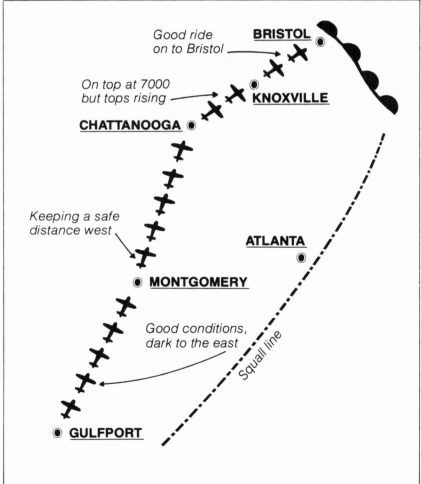

Good ride
on to Bristol ⟶ **BRISTOL**

On top at 7000
but tops rising ⟶ **KNOXVILLE**

CHATTANOOGA

Keeping a safe
distance west

ATLANTA

MONTGOMERY

Good conditions,
dark to the east

Squall line

GULFPORT

the low to be in Kansas, with a warm front to the east and a cold front to the southwest, and with the squall line all but outlining my proposed route of flight. I felt that the synopsis given to me was pretty much out of date. The whole thing was farther east.

The one advantage to something like a squall line is its well-defined nature. These things offer better possible/impossible choices than a lot of other weather situations. This one would surely pass my position before takeoff, and the best strategy would be to stay west of the line and hope it ended at some point. As far as I was concerned, that was both the best and the *only* strategy.

A last check of the weather at the airport drew the usual barrage of horror stories from the FSS. That is necessary, and even important; they have to make certain that pilots get the bad news as well as the good, but the tone of voice in which the bad is delivered is sometimes aggravating. The wind at Gulfport was out of the west, and the radar showed nothing to the north. All the activity was east and northeast. That was all I wanted—good enough weather to launch to the north, and get to where I could see and plan and talk with the controllers and other pilots. As long as I could see, the Sigmets weren't so important.

I discussed the situation with the ground controller. Even though the flight plan called for a route to the northeast right off the bat, I wanted to go north first, to avoid the weather. That was okay, and they furnished a clearance (out a radial of the Gulfport Vortac) that would be clear of all weather echoes observed on the traffic control radar.

After takeoff, it looked much better than the briefer had indicated. I was clear of cloud at 7,000 feet, above some lower cumulus; the dark, almost black, area to the east defined well the direction in which I should not point the nose of the little airplane.

The line of weather did appear as solid on their radar but some dissipation of actual thunderstorm cells was apparent to the eye. It was interesting to listen to other pilots pick around at the line. The Mobile weather was right at minimums, in heavy rain, and aircraft were shooting approaches and missing them for the most part. Others were penetrating the line, using a combination of information from airborne weather radar and the controller's information. The Stormscope in my airplane showed it to be pretty solid.

At one point, a pilot came through the line and reported heavy rain and moderate chop in his passage. The controller told another pilot about this and he asked to be vectored through the area from which the reporting pilot had just come. I guess he thought that heavy rain and moderate turbulence would be okay, but I hoped he understood the dynamic nature of storms and storm systems. What is heavy rain and moderate turbulence one minute can well be a deluge and severe turbulence the next. It's better to put stock in what it is now than in how it was a while ago. That your predecessor survived is no guarantee that you will.

I followed my planned route as far as Montgomery, but there it became obvious that something had to change. Atlanta, and its traffic and terminal control area, was ahead. If I continued as planned I'd be skirting the east edge of all that. The controller said there was a lot of weather up that way, and the thought of being sandwiched in between a TCA, a gaggle of jetliners, and a collection of thunderstorms made me opt to go west of Atlanta instead of east. There were no thunderstorms on that side, thus more options. I checked the weather over in the Carolinas and gave up on everything in that area as a fuel stop, too. It was all down around 200 feet, and the storms were still rumbling mightily.

My new destination became Bristol, Tennessee. The weather was good, and was forecast to remain so with only the chance of an occasional shower.

A solid layer of cloud beneath extended all the way from Montgomery past Knoxville. The tops built to above my level in the vicinity of Knoxville, but not long after I started through the tops of the bumpy clouds, they went away and revealed a beautiful picture of dark clouds to the east, freshly washed greenery below, and sparkling visibility when clear of clouds. The weather was adequate for a visual approach at Bristol.

One usual springtime storm ingredient was missing. The surface winds were not strong, and the winds aloft below 10,000 feet were quite light. I had only about 10 knots average tailwind on the northeastbound trip. This condition is often seen right behind a squall line, though, when the line of storms is more in relation to the upper air patterns than to the surface patterns. This day the surface low was off to the west, as was the cold front. The squall line was in connection with a jet stream moving around the southeast side of a low aloft and supporting the development of the line of storms. That hopefully meant that the line wouldn't be of indeterminate length and that I'd be able to proceed eastbound from Bristol and home without encountering any significant weather.

APRIL—BRISTOL TO TRENTON

North of the Squall Line

The briefer had good news for this flight, a continuation of the previous trip. There were no thunderstorms to the east, only rainshowers. The weather was good at Bristol, bad across Roanoke, Lynchburg, and Richmond, and then it turned good again in the Trenton area. The synopsis put the surface position of a warm front somewhere between Bristol and Roanoke. Winds aloft were forecast to increase from the southwest and then shift around to southerly on the other side of the warm front.

After takeoff, the view to the east was not a friendly one. The clouds appeared dark and layered, but as I drew closer the view moderated. Approaching Pulaski, I was on top at 7,000 and as the layers started to merge they did so rather gently. But there were still omens of bumpy clouds. Where the surface winds at Bristol had been light southwesterly, there were strong easterly surface winds over and to the east of the mountains. Those strong winds certainly raised the possibility of turbulence over the mountains, in addition to any turbulence that might be encountered flying through the slope of the warm front.

There was considerable turbulence just east of Lynchburg but it was strictly shear turbulence—no up- or downdrafts—and I related it to the frontal slope. Sure enough, after passing through the turbulence there was a substantial wind shift. Where I had been using little drift correction, I had to use 20 degrees to the right after flying through the bumpy area. Other pilots in the area were seeking information on the cloud tops,

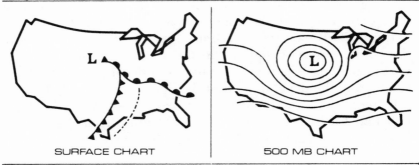

SURFACE CHART | 500 MB CHART

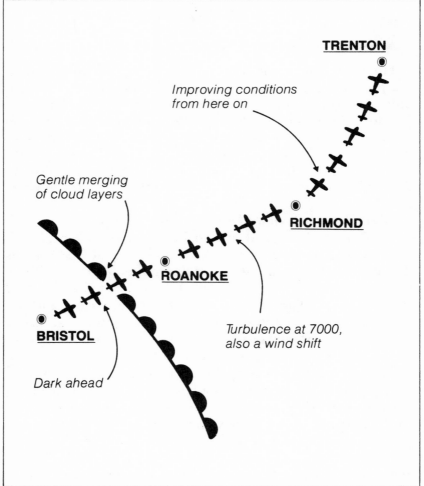

TRENTON

Improving conditions from here on

Gentle merging of cloud layers

RICHMOND

ROANOKE

Turbulence at 7000, also a wind shift

BRISTOL

Dark ahead

hoping for smooth air, but an air carrier reported solid clouds up to at least 12,000 feet.

Conditions improved in the vicinity of Richmond, the air turned smooth, and I was soon past everything but a few lingering (and slightly bumpy) showers. The wind moved around even more, probably to the south-southeast as the groundspeed dropped steadily down to 125 knots. There was a very fresh southeasterly wind on landing at Trenton.

The weather encountered on two flights this day, from Gulfport to Bristol and from Bristol on home, gives a good picture of the influence of the upper wind patterns.

There was a jet stream moving around the bottom of the closed low over northeastern Kansas, with 100 knot winds at the 18,000 foot level just south of the low and 80 knot winds to the southeast of the low. The circulation in the jet was a strong factor in the development of the squall line activity in the south, where the southerly surface circulation caused by the surface low to the west was providing an abundance of moisture. Further north, both the moisture supply and influence of the jet stream waned, the squall line diminished, and the last leg of the flight was through a rather benign warm front.

APRIL—TRENTON TO GREENSBORO

Strong Spring Low

There was a low pressure center over eastern North Carolina, said to be moving northeastward at 15 knots. The weather in New Jersey was good, and it was forecast to be good in Atlanta, the day's final destination, after some early day cloudiness. In between, well, it appeared wet and bumpy. A lot of rain was hosing Virginia and North Carolina, and there were Sigmets for ice and for turbulence below 12,000 feet. The winds aloft were forecast from the east at about 20, becoming northerly at 35 later on. These velocities covered all levels up to 12,000 feet. East winds along the east coast do not tend to increase so much with altitude, so this was logical. So was the turbulence. The radar summary showed 5/10ths coverage of rain and thunderstorms over eastern North Carolina.

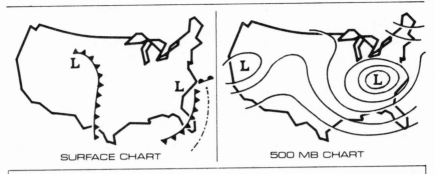

SURFACE CHART   500 MB CHART

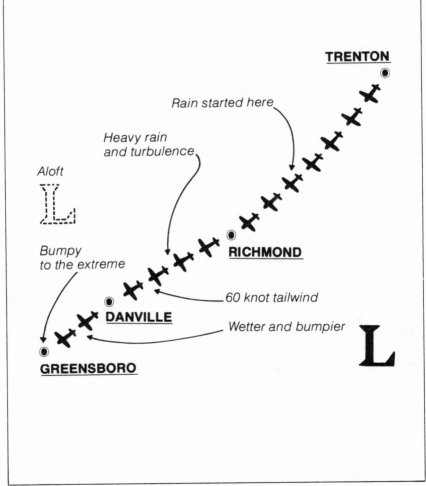

TRENTON

Rain started here

*Heavy rain
and turbulence*

Aloft

L

*Bumpy
to the extreme*

RICHMOND

60 knot tailwind

DANVILLE

Wetter and bumpier

L

GREENSBORO

The groundspeed wasn't super to start with, but at 15 knots over the true airspeed it was nothing to complain about. There were high clouds, and in the climb an inversion at 5,500 feet resulted in a temperature jump from +1 C to +5 C. Light rain started before Patuxent and by the time I reached that point it was wet and bumpy. The altimeter setting was dropping like a stone. The wind aloft at 6,000 was southeasterly to easterly, and the surface wind at Richmond was from the northeast at 15 to 30 knots. The wind and turbulence in cloud at 6,000 prompted a request for 4,000. The groundspeed increased at that level and the air was smooth between layers. The sun was dimly visible through a high overcast at times and, in all, I thought I had it by the tail.

The rain was heavier to the southwest, and the temperature dropped to 0 C at 4,000. I went down to 3,000, no alternative, because airplanes at higher latitudes were reporting ice.

The controller reported some echoes ahead but said airplanes had been going through. Did it ever rain: heavy at times, with light to moderate turbulence. It was shear type stuff, no up- and downdrafts, but very uncomfortable. The controller did say that aircraft were shooting contact approaches at Danville ahead, and an airplane in front of me reported that things were better over and southwest of Danville.

Conditions improved before Danville, but a bit past there the weather turned worse again. Heavy rain, turbulence, and at one point precipitation static built to the point that the radios were all but aced. The groundspeed on the Richmond to South Boston leg was 178 knots and I was running at about 120 knots true because of the turbulence. There was a stiff breeze blowing at 3,000 feet. Some tailwind: almost 60 knots.

Greensboro was reporting VFR conditions—1,400 overcast and ten miles visibility—but the turbulence in cloud at 3,000 feet was moving past moderate. There was vertical activity; the airplane was creaking and the flying was becoming ever more difficult. When the controller announced that there was an area of heavy precipitation ahead, just to the southwest of Greensboro, I felt that the present condition was uncomfortable enough for me, and that I should pause for refreshment at Greensboro. The weather the controller saw made enough of a mark on radar for all radar-equipped flights to deviate around it, and the Stormscope in my airplane was showing a rather general electrical pattern ahead. All the more reason to rest a while.

The air was so rough in the approach at Greensboro that my Jeppesen book took a little trip around the cabin and wound up on the floor. It smoothed out a bit below the clouds, though, and I nearly fell into one of the oldest sucker traps. The visibility was good and the ceiling was high; why not just head southwest, VFR? I mentioned that to the controller who simply said that rain was moving toward the field from the northeast, and that the area to the southwest looked solid. I landed at Greensboro.

Checking the weather, I found a great collection of new Sigmets about turbulence, any one of which I could verify. But the implication was that the turbulence was caused by high winds over rough terrain, and I *knew* they were in error on that. The winds aloft were easterly to northeasterly, and from that direction they wouldn't be blowing over high terrain at all. The bumps were from something going on in the weather system itself.

After about forty minutes, a check of the weather radar showed no thunderstorm activity to the southwest, so I launched again, for Atlanta.

It's amazing how markedly weather can change in a short period of time. There was one shower to go around, vectors provided by the controller, but basically the ride to Atlanta was a smooth one at 4,000 feet. This further reinforced my feeling that the turbulence was caused by some development to the west of the low pressure center, not by wind over high terrain. That is contrary to the principle that the weather improves when the low is to the east of your position, but the 500 mb chart tells the tale. A closed low aloft was back to the west of the surface low so the swirl was in that direction, into the upper low, and I didn't find better weather until I was west of the upper as well as the surface low.

APRIL—ATLANTA TO GAITHERSBURG

Once Again, in the Other Direction

I watched the evening weather on the TV after the preceding flight and they reported from three to four inches of rain over much of North Carolina. I sure believed it, and the next morning I hoped that the low had moved away, to let me have a nice ride back northeastward. When the time came it was still a factor, though. The low was a bit east of its position of the day before, so the weather was better, but the headwinds would be there in spades: northeasterly at 30 to 35 knots, same direction and velocity up to 12,000 feet. The weather was reported and forecast to be okay, with 3,000 to 5,000 broken predicted over the area where I had been in the rain and turbulence the day before.

The climb was in smooth air and the ride good after leaving Atlanta. The groundspeed was as expected at our 9,000 foot cruising level: 103 on one leg and 108 on the next, giving the forecaster pretty good marks for his 30 to 35 knot forecast as applied to my 140 knot airplane.

I came on a cloud layer around Greensboro, and started picking up a trace of ice in clouds. Most of the time, though, I was in snow which became progressively heavier but didn't do much to ice the airplane. The static electricity build-up was quite spectacular, and any time I put my hand within a foot of the windshield it would get an enthusiastic spark.

The air remained smooth for the most part, but there were some jiggles as the airplane passed under a rather dark cloud and flew into what seemed a different situation—heavier snow and cumulus type clouds building into the flight level. It *was* different, too. The groundspeed dropped to as low as 85 knots in this area, indicating a wind of 55 knots out of the northeast. I tried lower altitudes, all the way down to 3,000 feet, but with no improvement in groundspeed. The 3,000 level was handy for melting the ice that had accumulated, but it was so rough that I went back up as soon as all the frozen stuff was gone. Before reaching Gaithersburg, I flew into an area of unrestricted visibility beneath a high overcast.

That change in weather en route and drop in groundspeed

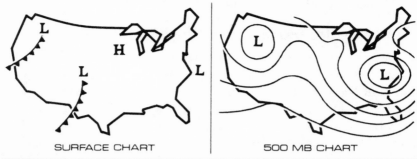

SURFACE CHART   500 MB CHART

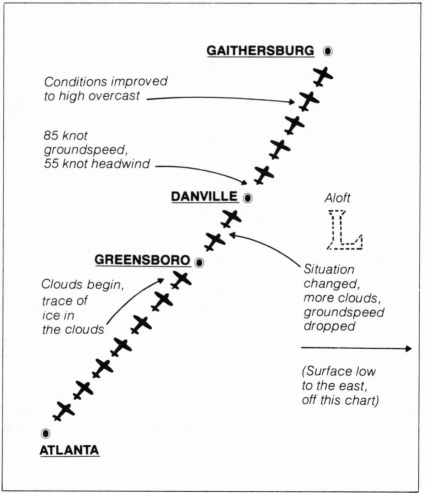

GAITHERSBURG ⊙

Conditions improved
to high overcast

85 knot
groundspeed,
55 knot headwind

DANVILLE ⊙   *Aloft*

                                    L

GREENSBORO ⊙

Clouds begin,   *Situation*
trace of   *changed,*
ice in   *more clouds,*
the clouds   *groundspeed*
                                   dropped

                                   (Surface low
                                   to the east,
                                   off this chart)

⊙
ATLANTA

was in the same position relative to the surface low as the higher tailwind and wet and bumpy ride of the day before. There have been other instances of being bumped around to the west of a surface low, and in each case there was probably the same relationship with a closed upper level low.

When there is a closed low aloft, the movement of the surface low is often slower. Under certain conditions a surface low will even move back to the west and position itself under a low aloft but this one didn't quite follow that pattern. However, it did move slowly; it became very strong and spawned high winds and record rainfalls, and the weather was about as bad to the west of the surface low as to the east. There was development and building to the west: While the circulation was around the surface low, much of the upward action was into the low aloft.

Why the big burst of wind aloft on the west side of the surface low? The winds forecast made no suggestion that this would be the case, so there is apparently nothing in the meteorological computer that suggests such a variation in wind on the backside of a surface low, regardless of upper air conditions. It would seem logical, though, that it was at this point that the maximum circulation into the upper low was encountered, causing a general acceleration in wind velocity.

MAY—TERRE HAUTE TO WICHITA

Developing System

The briefing was in the Terre Haute FSS, and the specialist read off the list of Sigmets and Airmets in his dourest voice. The basis of the trouble was a slow moving low in western Kansas, and the advisories called for strong low level winds and turbulence plus embedded thunderstorms to the east of the low. A pilot report told of an east-west line of storms forty miles southwest of Butler, Missouri, which was along the proposed route of flight. The weather everywhere was good— 4,000 to 5,000 broken with a higher overcast and with some scattered rainshowers over a wide area. Forecast winds were on the nose, naturally, at about 30 knots. I spent a few minutes looking at the 500 mb chart on the wall, and the pattern indicated that there shouldn't be any severe activity, for a while

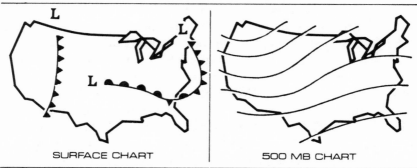

SURFACE CHART   500 MB CHART

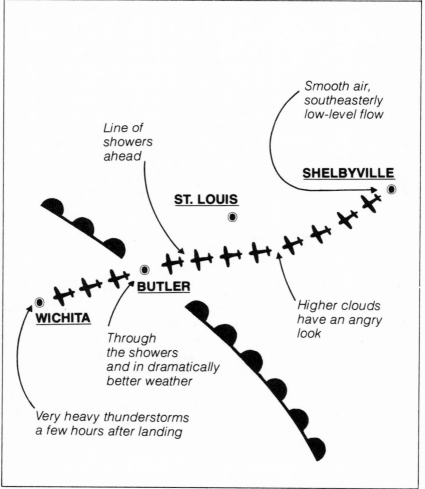

Smooth air,
southeasterly
low-level flow

Line of
showers
ahead

SHELBYVILLE

ST. LOUIS

BUTLER

Higher clouds
have an angry
look

WICHITA

Through
the showers
and in dramatically
better weather

Very heavy thunderstorms
a few hours after landing

at least. A trough was developing to the west, over the Rockies, and there was a southward protrusion of the −15 C temperature line that headed back north over western Kansas. The surface temperature in Wichita was +32 C so that colder air aloft could cause problems toward the end of the flight. The temperature drop with altitude would be well over 2.5 degrees C per 1,000 feet, a good sign of instability.

I flew for a while at 4,000 and then at 6,000, trying to minimize the headwind as well as stay below the bases of the clouds. It was rather smooth down low, and pilots flying above the bases reported a rough ride. The air at lower altitude was stable enough, with the temperature at 6,000 only about 8 degrees C cooler than at the surface. The map didn't show a warm front, but as is often the case, there was something east of the low resembling one.

Some rainshowers appeared when I was not far southwest of Terre Haute, but there was no scud forming in these rain areas because of the relatively warm temperature at the surface. There were even thunderstorms, and at one point I flew around what appeared to be a rather light shower that was producing spectacular cloud-to-ground lightning. The air was definitely unstable aloft.

As an aside, there is really nothing prettier than springtime after a bath, as seen from an airplane. The greens are delicate, and as I moved along dodging the scattered showers I was getting quite a visual treat.

The controller said that the rain activity ended about fifty miles west of St. Louis and indeed it did. The altimeter setting was dropping like a rock, though, and if there was to be any "big" weather it would be to the west.

As I was passing Butler, the controller was telling pilots of thunderstorms forming on a line from just south of Salina, Kansas, up to Omaha, Nebraska. Nothing was showing south of Salina but, as the controller said, "Those things can develop rather quickly." The Wichita weather remained good, and the view from the airplane was clear with good visibility—no way for nature to hide anything in conditions like that.

I never saw anything resembling a line of activity around Butler, or anywhere else east of Wichita, as had been indicated in the pilot report received at Terre Haute. Pilot reports are really of very little value in thunderstorm areas if they are more than twenty or thirty minutes old because conditions change too fast.

When I reached Wichita, the forecast southerly surface wind was reported as around to southwest; as I made my approach I saw that the windsock was even favoring the west-northwest. Had a front passed the area? Something was surely having an effect as there was quite a build-up to the southeast of Wichita that extended into a developing line over Oklahoma, a southerly adjunct of the activity up around Salina. Within an hour after I landed, reports of tornados on the ground to the south and east of Wichita were coming in.

After landing, I went back out for a local flight in another airplane and flew toward the building storms to the southeast just to have a look at the development. It was quite rapid—you could almost see the things popping—and later on in the motel I had a look at the local TV station's color radar. The storms had developed into a broken line, some miles to the east of Wichita, and they appeared quite strong. I went to bed that evening thinking that the storms to the east and the westerly surface wind meant that the front had passed through Wichita and that the next morning would dawn bright and clear.

MAY—WICHITA TO CINCINNATI

It Happens Every Spring

This, the day following the preceding flight, was full of surprises. The first was a dense fog. The front had passed the afternoon before, right? And after a cold frontal passage you don't expect a dense fog the next morning, right? Wrong. At least the cold front didn't remain in its position of the afternoon before, because as I was peering out the motel window at the fog I saw a banner on a pizza parlor across the street indicating a southeast wind. I should have expected something like this. Remember, there was no close-by trough on the 500 mb chart the day before so the low didn't have any strong moving and steering influence aloft. I had made a little move the afternoon before and had then stopped for the evening and perhaps even backed up a bit.

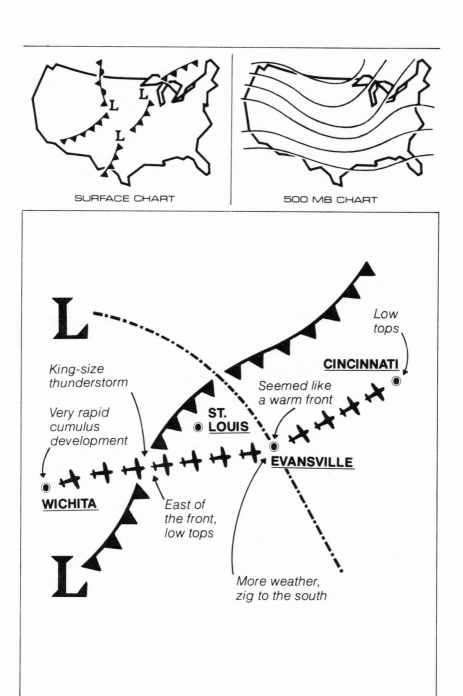

SURFACE CHART

500 MB CHART

L

Low
tops

King-size
thunderstorm

CINCINNATI

Very rapid
cumulus
development

Seemed like
a warm front

ST.
LOUIS

EVANSVILLE

WICHITA

East of
the front,
low tops

L

More weather,
zig to the south

The man at the flight service station was rather vague. He said there was a cold front in central Kansas, apparently west of Wichita, and that there were a lot of low ceilings around. Wichita was reporting sky obscured and an eighth of a mile visibility. Kansas City was in similar shape. To the east, toward St. Louis, the weather was good. Radar reports showed activity in north-central Kansas and in the Ohio Valley. The briefer suggested that some of the thunderstorm activity wasn't too far away and that I should check conditions right before takeoff. My destination for the first cross-country leg of the day was Dayton, for a fuel stop, but the briefer didn't have and couldn't get Dayton weather or a forecast. I compromised and selected Columbus as a destination because their forecast was good (5,000 broken, 10,000 overcast) and the forecast winds were adequate to push me to Columbus with fuel to spare.

The "Today" program weather map was a bit more concise than the FSS briefing. There was a low just about over Wichita with a front to the northeast and a front to the southwest. The area to the east of the low and the fronts was slated for severe thunderstorm activity during the day.

I was to get an even better briefing for Wichita weather because I had a local flight in another airplane scheduled prior to my eastbound departure.

Another pilot checked the weather for that local flight, and told me that tops were 3,300, clear above, and that we'd simply get a clearance to VFR on top, do the flying per the program, and then shoot an approach back into Beech Field at Wichita. The ceiling was reported at 700 feet, visibility five miles. The surface wind was still southeasterly. Typical low stratus of a springtime morning.

At the runup pad for a south departure, I checked the windsock. It showed a northwest wind. The tower confirmed this and told us to use the north runway. I asked how the visibility to the northwest looked, and the tower operator said it was fine, he could see downtown clearly.

The clearance was to 4,000 feet, to maintain four and advise if not VFR on top at that altitude. Right after takeoff, I knew we wouldn't be on top at any reasonable altitude because it started raining, lightly at first and then more heavily. There was some light turbulence, too, indicative not of early morning stratus, but of rainy cumulus clouds. That type doesn't top at 3,300 feet.

It didn't take any weather-wise genius to figure out that the flight was being conducted in the cold frontal zone. There still didn't seem to be much to fret over, though, because the controller said that he wasn't painting any weather except to the north. I was southeast bound.

Up to six and still not on top, but at 7,500 we did break out. It was quite a sight, too. Instead of actually flying out on top, it was more a matter of flying out the side. To the southeast lay the stratus and the low tops. To the northwest the tops were much higher, were growing, and there was a big cumulonimbus to the north. The cold front had some life to it.

After completing the mission in this airplane, which took only a few minutes, it was time for the approach. If indeed that was a frontal passage a while ago, this approach would be back through the front. The nature of the beast was beginning to show, too. It was clearly a cold front with Wichita surface winds around out of the north at 30 knots.

After carefully eyeballing the approach course before starting into the area of cumulus, I thought that it looked pretty good. The bad stuff was to the north, and the approach would be in an area where the tops looked not over, say, 12,000 feet. The controller said that he wasn't painting any weather in the area of interest, so I felt pretty good about the whole thing.

The approach went well except for some brief, very heavy rain and some turbulence in what must have been the area of transition from the southeast to the strong northerly wind. An air carrier reported severe wind shear on approach to Wichita Municipal, but that was more of a problem to his heavy airplane than mine. The light airplane handles wind shear better because it can adjust to new situations more quickly.

The landing was into a very strong northerly breeze. I was ready to go again in about an hour, and the FSS briefer reported that there was a strong storm developing rapidly to the northeast of Wichita but that nothing was shown to the east or the southeast.

The normal route was to the northeast, but right after takeoff I told the controller I'd like to head around the south side of that big thunderstorm that was getting bigger by the minute, and he approved a heading of 120 degrees.

I was on top of most clouds at 7,000, but the occasional ones I was flying through were quite bumpy so I asked for 9,000. It was a little better there, but cloud tops were building rapidly and 9,000 was soon a very turbulent level. I asked for 11,000. At that lofty perch (lofty to a light airplane pilot, anyway) I could clearly see the limits of the storm, and could plot a path to the southeast. It worked pretty well; I busted through the tops of only a few cumulus, but visually the situation was as chaotic as I've seen. The growth of the cumulus was almost explosive.

About sixty miles east of Wichita I got back ahead of the front and over that peaceful low stratus. It was nice to have that behind, for it would surely spawn a real squall line later in the day. The benefits of flying in the morning are great in thunderstorm season.

Not long after I was free of the front, a fresh Sigmet suggested that the thunderstorm to the north was quite something: tops to 48,000, hail, surface wind gusts to 80 knots.

It's not unusual to feel that all is behind you after tangling with a front three times in one day, but there was to be further entertainment. Over Vichy, Missouri, I could see a substantial build-up ahead. In answer to my question, the controller said that it was east of Centralia, which was 124 miles ahead. When you can see one that far away, it is pretty big.

I amused myself for the next thirty minutes trying to decide whether to go north or south of the storm. The controller suggested that my route as filed would take me to the north of it, but there appeared to be a congested area of building cumulus to the north, and I estimated their tops at well over my 11,000 foot cruising level. To the south looked better because there didn't seem to be any cumulus growing on that side of the storm. I opted for a southerly route, changed my flight plan, and flew happily toward Evansville.

The controller was correct when he suggested going north of the thunderstorm. At least his way around couldn't have been a lot worse than my way. At 11,000 I moved over an area of rapidly building cumulus and then found myself in a real thicket, building tops and no clear path to follow.

I've long had a theory that in such a situation the best way out is down, to a lower altitude. This was reinforced by the Evansville weather, 1,000 scattered, 4,500 overcast, three miles in light rain. If I could be below the cloud bases, it would be smoother. A descent was approved, and I picked a heading that would be toward the visible area with the lowest tops and started down. The ceiling was higher than reported because at 5,000 I was beneath the bases of the clouds. There was some heavy rain for a while but not much turbulence. Then I flew into another weather situation: On top at 5,000, I had to go up to 7,000 to stay on top. There the air was smooth, with no build-ups apparent ahead. Ahh, surely I had it made. Instead of pressing on to Columbus, I landed at Cincinnati because I was ready for a soda pop. I still had to continue to Trenton, N. J., but Cincinnati was better than halfway, and I anticipated an easy flight home. That weather back around Evansville, I thought, was connected with something resembling a warm front. No front was depicted on the chart but if you had drawn one to the southeast from the low that was north of Wichita, along a logical path, that's about where it would have been.

MAY—CINCINNATI TO TRENTON

When Least Expected

This flight was a continuation of the previous trip, and I was pleased when I listened to the briefer go through a breezy "no problem" review of a "high pressure ridge off the coast" and the weather ahead. "Just not much there," he said. "About the only thing you'll find is a bit of light rain over western Pennsylvania. I can't guarantee it, but you'll probably be between layers all the way at 7,000 feet." When he read the Columbus weather, 900 overcast and five miles in rain, I couldn't help but recall the forecast of excellent weather at Columbus when the day had started. But then it had been a long day and things do have a habit of changing. Winds aloft were to be southwesterly at 30 knots all the way.

I was on top after leaving Cincinnati. It was a nice stable situation with pretty cloud formations, but that wasn't to last. About forty miles east of Columbus I started running into

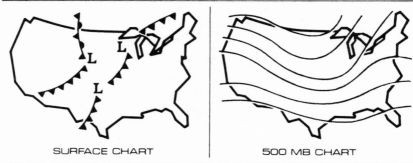

SURFACE CHART   500 MB CHART

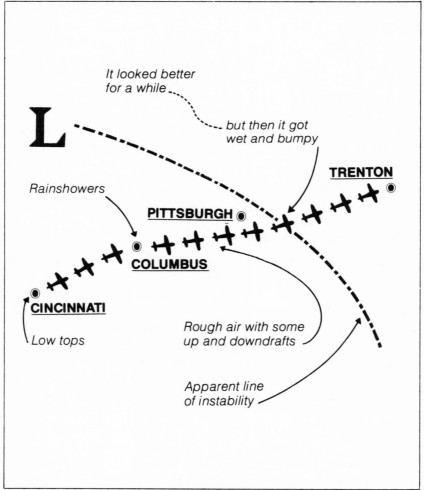

It looked better
for a while

but then it got
wet and bumpy

L

Rainshowers

PITTSBURGH

TRENTON

COLUMBUS

CINCINNATI

Low tops

Rough air with some
up and downdrafts

Apparent line
of instability

some rainshowers, and wound up in cloud at 7,000 feet. Still, things were not too bad and I was flying toward better weather.

Nearer Pittsburgh, I was beneath the bases of a basic cloud formation and it became very rough with some moderate up- and downdrafts. There were rainshowers around, and there were some wisps of clouds beneath the basic overcast that suggested vertical action. As best I could tell from industrial smoke, the surface wind was from the west, where it was supposed to be southerly. The winds aloft were strong and southerly instead of southwesterly as I was checking a 10 knot headwind component on an easterly track.

I felt like 5,000 might be a more comfortable altitude because it would be farther below the cloud bases, but the controller rather curtly answered my 5,000 foot request with "negative." Then things seemed to get a little better at 7,000. The visibility improved, the turbulence subsided for a bit, and once again I thought that the weather was behind. The controller called and offered 5,000, in a pleasant voice this time, and I declined, saying that things were smooth at 7,000 now.

Closer to Johnstown, I began to wonder. It looked dark ahead, and in a flash I learned that 7,000 would no longer be beneath the bases of clouds. I was entering an area of cumulus with much lower bases, and after making one stab at a deviation to the north I settled in to fly straight through. I asked the controller if he was painting anything and if I could still have 5,000; the answer to both questions was negative. Next came a very heavy rainshower and enough updraft to make me have to extend the landing gear to keep the altitude near 7,000 feet with the airspeed on a proper value and the engine developing enough power to stay warm. I repeated to the controller that I'd like 5,000, and reported moderate turbulence at seven. This prompted action on his part and he said I could have five if I'd be there in two minutes. You bet I would.

By the time I reached five, I had flown out of the east side of the rainshower. I looked back over my shoulder and the area looked even darker from the east side than it had from the west.

Finally that elusive dream about having all the weather behind came true, and the flight into Trenton was smooth and clear of clouds.

I guess we get too soon old and too late smart, for on this run I was again spoofed by the same thing that had been a factor on the prior flight. There was no feature on the map, but there was a low to the northwest, and inclement weather was found right about where a warm front would logically have been found on the map if all the frontal qualifications had been met. And where the action in the area was not enough to attract widespread attention, it was adequate to add spice to a low altitude flight in a light airplane. In this instance, the spice was enhanced by the mountains. A southwesterly flow to the west and a southeasterly flow to the east also meant convergence with some lifting enhanced by terrain to total up to the showery weather.

The wind aloft being stronger and more southerly than forecast was an indication of more than just some light showers, too, and should have been an early tipoff to the possibility of a little action.

Whatever the cause, the leg that looked easiest turned out to be the bumpiest. Earlier, around Wichita, and later in Indiana, I had been dealing with bigger weather that was easily identifiable. The weather close to Wichita was big indeed; it spawned 25 tornados in states I flew over that day. This is the type of weather that briefers become hyper about. But if we keep an eye out and choose paths carefully, it can be less of a problem than some situations not specifically identified by the human or computerized forecaster.

MAY—PHILADELPHIA TO CHARLESTON

No Cold Front?

The weather had been rainy in the northeast for almost a week, but the morning of the flight held promise for the end of the wet stuff. There was no dramatic clearing predicted, just a slow evolution to warm and hazy sunshine. The synopsis positioned a weak low over northern New York State and another one off the coast of New Jersey. The latter low had been there all week, making rain, and was weakening rather than moving. The trip to Charleston was projected to be a good one: winds aloft rather light and westerly, no echoes on radar, and a forecast of clear skies for our arrival. Philadelphia was to be 4,000 broken, variable to scattered, with five miles visibility.

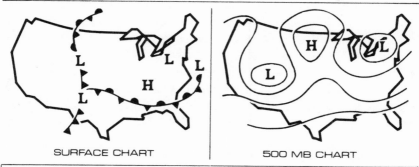

SURFACE CHART

500 MB CHART

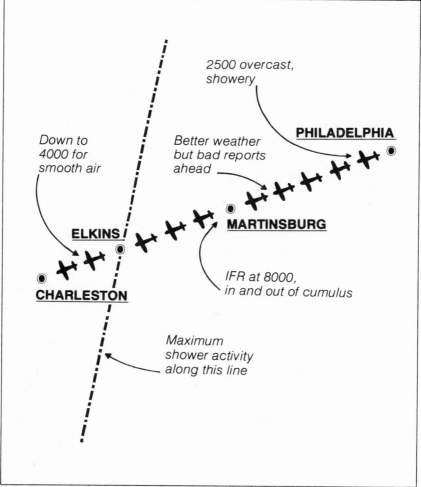

2500 overcast,
showery

PHILADELPHIA

Down to
4000 for
smooth air

Better weather
but bad reports
ahead

ELKINS

MARTINSBURG

CHARLESTON

IFR at 8000,
in and out of cumulus

Maximum
shower activity
along this line

They didn't have the IFR flight plan that I had filed at Philly, so I departed VFR and headed west, to pick up the clearance later. At times this looked like a mistake. There were rainshowers around, the ceiling was more like 2,500 overcast, variable to broken, than 4,000 broken, variable to scattered, and there were moments of picking between various shades of gray to determine the best heading to fly. The flight continued VFR, though, because they still couldn't find the IFR flight plan that had been filed.

The situation brightened some when close to Martinsburg, West Virginia, but a check of the en route weather suggested that it would be prudent to start all over, file a new IFR, and get a clearance before starting across the mountains. Elkins, West Virginia, was reporting 2,000 overcast and four miles in light rain which strongly suggested that VFR across the Appalachians would be uncomfortable, if not impossible.

The clearance was at 6,000, then on up to 8,000 feet. I had originally thought we'd be on top at 8,000 feet but it was solid at that altitude except for times when we'd break out into canyons of cumulus. These weren't building rapidly but they were a little bumpy and there were some fairly enthusiastic showers around. I wandered around a little, aiming at the lightest spots, but it didn't seem to matter that much.

Charleston was reporting 4,000 overcast and nine miles visibility, so once Elkins was behind I requested 4,000, to get below the bases of the clouds. That helped. The air was smooth and the ride on in was serene.

The weather was no real factor for an IFR flight but it could have been touchy for a VFR flight even though the information available at the FSS before flight suggested no problem for visual flying. The worst forecast along the route was 3,500 scattered, and on a depiction chart that one of my passengers had procured through his company's private weather service, the word was to expect only layered clouds at 8,000 feet. They did show some scattered showers, tops to 20,000 feet, just north of our route. Apparently the showers and poorer than forecast weather at Elkins was an unforecast southerly extension of this activity.

The weather encountered was a product of that low over northern New York in combination with a closed low aloft right above it, with a trough aloft to the southwest. Even though I didn't have the 500 mb chart, or any description of it, something should have been anticipated by projecting possible frontal positions based on the position of the surface low. The weatherfolk were not drawing fronts on their map but in this case it was the cold front that wasn't there, plus mild action encouraged by the mountains, that caused the showery weather. It was almost the same deal as the preceding flight where a condition similar to a warm front created a wet and bumpy situation over the mountains.

JUNE—BRISTOL TO LITTLE ROCK

Weak Front

There was a cold front on the map, aligned northeast-southwest and draped across Tennessee. The briefer said that a line of scattered thunderstorm activity was showing on radar in the frontal zone but that no stations were reporting activity. The reported weather was good, with Jackson, Tennessee's 900 scattered and five miles the worst thing going. The winds aloft were forecast to be light southwesterly.

The surface temperature was +26 C and at 6,000 it was +17 C so there was no apparent instability in the lower levels. I didn't check the wind and temperature forecasts for 18,000, which was an oversight. That information would have been useful in judging the possibilities for convective activity as I eyed some cumulus building through 6,000 just west of Hinch Mountain in Tennessee.

The cumulus weren't particularly bumpy, but the sky ahead did appear dark. The Stormscope installed in the airplane displayed no electrical activity, though, and the controller said that all she painted were some scattered rainshowers around Shelbyville. Other aircraft had been moving through the area without complaint, too, so everything seemed on the up and up.

The rain was falling from a higher overcast, and when I reached the precipitation area there weren't any cumulus around—just the high clouds and rather gentle rain. It took but a few minutes to get through it, and the ride was actually

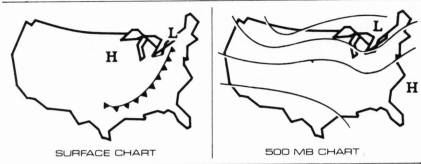

SURFACE CHART   500 MB CHART

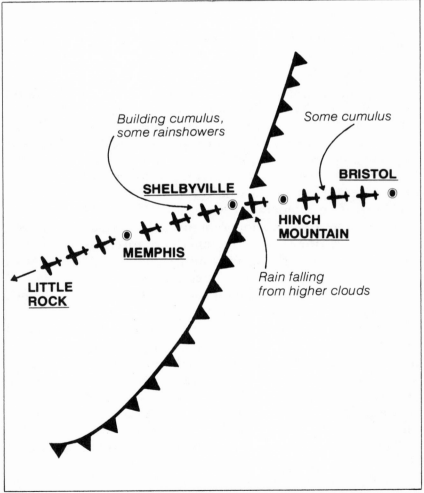

Building cumulus, some rainshowers

Some cumulus

BRISTOL

SHELBYVILLE

HINCH MOUNTAIN

MEMPHIS

Rain falling from higher clouds

LITTLE ROCK

roughest after the controller said that I was west of all the weather. There were more cumulus to the west, and some had built into rather bumpy rainshowers, tops probably around 15,000 feet. There were no build-ups higher than that, though, and it was relatively easy to dodge most of the clouds and showers. The higher clouds had gone away and the weather looked more typical of warm sector weather in the tropics than post cold frontal weather. There was no temperature change behind the front, and the barometer didn't move up substantially.

In the very late spring and the summertime, cold fronts do often peter out and become almost unidentifiable, as this one did. Given no abundance of moisture or instability, they just can't create much weather. Too, this flight was flown rather early in the morning so there had been no heating to aid and abet development in the frontal zone.

JUNE—DALLAS/FORT WORTH TO WICHITA

Severe Thunderstorms

The first weather information for this morning came from the "Today" program map. It featured a front between DFW and Wichita, turning from a cold to a warm front in that area, with strong thunderstorms indicated north of the front. Those storms were on the radar early in the morning, and they showed on the "Today" satellite picture as a bright blob of clouds in the area north of Dallas. The NWS map, obtained later, showed the front as stationary.

The day's flying actually started in Kerrville, Texas, and the trip from there to DFW, to drop a passenger, was uneventful. The DFW weather was sparkling clear on arrival. But on checking for the continuation, I found that all the early morning indications of bad weather were coming true. There was an east-west line, 40 percent coverage, and thunderstorms, some severe, north of Oklahoma City. Tulsa and Ponca City were reporting thunderstorms, Wichita's storm had ended at 35 past the last hour. A convective Sigmet covered the area and called for occasional severe storms, with tops over 50,000 feet. Two things were clear: One, it was certainly reasonable to start the

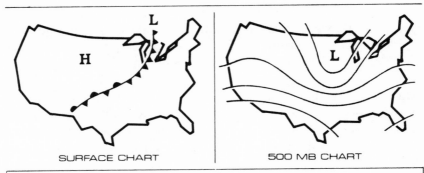

SURFACE CHART

500 MB CHART

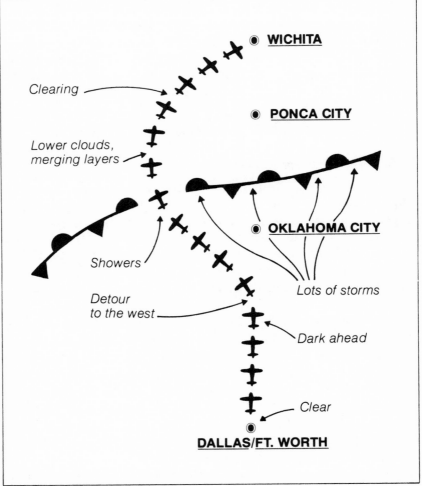

⊙ **WICHITA**

Clearing

⊙ **PONCA CITY**

Lower clouds,
merging layers

Showers

⊙ **OKLAHOMA CITY**

Detour
to the west

Lots of storms

Dark ahead

Clear

⊙
DALLAS/FT. WORTH

flight and have a look. Two, it was equally certain that the flight plan wouldn't be engraved in stone. It was clearly one of those situations where you must remain flexible.

The Stormscope in my airplane and information from the controller both strongly suggested that there would not be any way to proceed as filed, over Oklahoma City and Ponca City. There was a *lot* of activity there, and the controller said that it was solid from 40 miles west of Oklahoma City to pretty far east of Tulsa. Further west of Oklahoma City looked pretty good to him, with the activity out that way more scattered.

The suggestion was to fly a heading of 270 degrees for about fifty miles and then head north. Using the Stormscope's indication, I refined that heading to 290 to minimize the amount of miles flown on the detour. It could always be adjusted back to 270 if necessary.

The storm area was just north of Oklahoma City and was very black in appearance. The controller said that it looked to him like one storm all the way over to Tulsa. In spots, the activity had a greenish tint that is sometimes associated with extremely heavy rain. It was maintaining a good distance from the storm, and while it didn't look to me like I was the twenty miles away that should be observed when severe storms are forecast, the controller said that I was twenty miles away from the heavy weather he was painting. Whatever, the appearance of the situation ahead was okay.

West of Oklahoma City I observed a hole through the weather and for a fleeting moment entertained the thought of ducking through there. It appeared bright on the other side of the hole, but there was something about it that suggested a true "sucker hole." Indeed, as I examined it more closely, I could see fractured and wispish clouds being tumbled about in the area. Rain was starting to fall in the hole, too, and it would have been a very bad deal. I don't know what was on the other side, but it was one of those situations where human vision is the best turbulence avoidance device going. The electronic devices might have suggested that a ride through there wouldn't be bad, but what I saw convinced me that detouring a few more miles would be better.

There were some developing showers to the west, with bases above my 7,000 foot cruising level. It did seem that bases of any developing activity would be higher than seven because the air was actually rather stable in the lower levels with a temperature drop of slightly less than 2 degrees C per thousand.

West of Enid, Oklahoma, I was on top of lower clouds at 7,000, with higher clouds above. It looked like the cloud layers would merge ahead, which they did, and my primary interest was in making sure there were no embedded cells in the clouds. The controller said that there was some scattered stuff along my route of flight, but he said that his radar was not accurate enough for vectors around weather. I descended to 5,000, in hopes of finding visual flight conditions there, and continued on the basis that the Stormscope showed no return directly ahead. The 5,000 level might be below the base of any shower activity, too, giving a better ride.

Conditions were off and on visual at five, but it didn't matter for long as I soon flew out into brilliant sunshine on the backside of the line of weather. The freshly washed wheat fields were a magnificent sight and the view lasted almost to Wichita. A lower overcast had formed there with bases 1,000 above the ground and tops about 1,000 feet higher.

The situation this day begs for an explanation because it ran counter to some of the basics of meteorology that we have discussed. There was not a strong low level southerly flow to feed moisture, and there was no substantial instability in the lower levels to the south of the front. Additionally, there was nothing going on at the 18,000 foot level to spawn severe storms. The low and trough on the 500 mb chart were to the east.

This was apparently a case of having just enough of each ingredient to make storms. There had been rain in the area for several days, so some moisture was coming from the ground as it was heated as well as from the 10 to 20 knot southerly flow. There was lifting from convergence in the frontal zone as the southerly wind bumped into a north to northeasterly wind north of the front. And there was instability aloft to support the whole thing. I would have known this had I checked the temperature at the 18,000 foot level. I surmised that there was no instability in the lower levels on the basis of a +20 C temperature reading at 7,000 and a +30 C reading on the surface, for 10 degrees of drop in about 6,000 feet for that area of elevation. At 18,000, though, the temperature was −10 C for a 30 degree drop in the 11,000 feet above 7,000, which is almost 3 degrees per thousand feet. That's plenty. The weak front provided enough lifting to get it going, and once above 7,000 feet the instability provided the necessary support.

JUNE—WICHITA TO CINCINNATI

We Meet Again

This flight was the day after the preceding one and the weather map offered on the "Today" program was much like the previous day's with one important exception: There was now a low pressure center positioned just to the west of Wichita. The forecast thunderstorm activity was to the east of the low, reflecting the standard pattern. The reported weather at all stations was good but radar was showing a broken line of activity along the Kansas/Missouri border, from Topeka to Joplin.

There were some pretty big storms to the south of Wichita, but nothing to the east or northeast to begin with. There were altocumulus clouds, based at about 10,000 feet, and some of these appeared to be building rather rapidly. My altitude was 9,000, and the temperature there was indicative of a rather stable layer of air up to 9,000 feet.

The north-south line of weather was about forty miles east of Butler and aircraft were going through with only slight deviations. The Stormscope showed some electrical activity, but not a lot; it and the traffic controller agreed that a slight deviation to the south would handle things. The general appearance of the situation strongly supported the radar and Stormscope's observations.

The controller said that the line was about twenty-five miles through, and I could easily see the other side. I could also see from the cloud bases that things above 10,000 were probably quite active. The clouds were churning on the bottom, they were very dark in color, and in spots they had a mammatocumulus appearance. I expected some turbulence, and found some there. There was an area of updraft, then of downdraft, with some shear turbulence in between. All the various

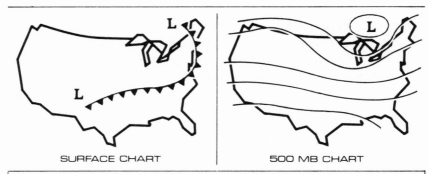

SURFACE CHART | 500 MB CHART

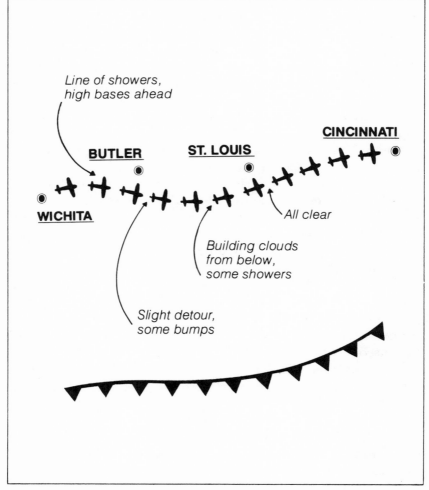

*Line of showers,
high bases ahead*

BUTLER   **ST. LOUIS**   **CINCINNATI**

WICHITA

All clear

*Building clouds
from below,
some showers*

*Slight detour,
some bumps*

stages of convective activity were there. I was just below the majority of the action, in more stable air, and I sampled only a mild version of each.

The temperature at 9,000 dropped by more than 5 degrees as I passed under the cloud. There were lower clouds ahead with building tops, and I was soon in cloud at 9,000 feet. I saw some build-ups to about 15,000 before entering the clouds, and encountered some bumpy and showery weather for about the next 100 miles. Then I flew out into nice clear weather. There was a lot of activity to the south all along, and aircraft going in that direction were making lengthy diversions around weather.

This was almost a carbon of the day before, only it was a little earlier in the day before heating had become as much of a factor in encouraging development. The temperature at 18,000 was still − 10 C, and there was instability from about 10,000 on up but a relatively stable layer below that.

The effect of temperature changes on cloud formations can clearly be seen in what happened after I flew under the line of showers. The temperature at 9,000 dropped, but the surface temperature remained the same. As a result, there was more instability in the lower levels and, sure enough, lower clouds started building through my altitude as proof of that simple fact. The front was shown on all charts as being south of my route, and indeed most of the activity was in an east-west line and in the charted location of the front. The north-south activity around Butler must have been related to a weak wave on that front, or to a weak trough of low pressure in that area.

Flights: Third Quarter
July, August, September

July is the month of haze and thunderstorms. A stationary high pressure center off the east coast, called the Bermuda High, can result in a prolonged period of warm and moist southerly flow over the eastern U. S., with haze tops well above 10,000 feet and with occasional thunderstorms embedded in the haze. But for the most part July is a fun month in which to fly, the ultimate time for vacationing and outdoor activity.

By August, the weather starts exhibiting signs of change. The shorter days cut average temperatures, and cold fronts, though rather mild, begin pushing south. If June is the nicest month of the year in which to fly, August is the second nicest.

September is thought of as being the heart of hurricane season. A hurricane (an intense low pressure area) tends to cause good weather everywhere except in the storm itself. Position one out in the Atlantic several hundred miles, and it'll take in all the moisture for miles around and leave the east sparkling clear. But watch out if it turns toward you. Flee, and do so well in advance. A hurricane can be more than a match for the best of tie-downs, and strong ones can wipe out the sturdiest hangars.

Toward the end of September, the nice fall weather starts in earnest. Cold fronts push farther south, and the weather associated with high pressure areas takes on a new clarity. No question, the weather is ready for a major change.

JULY—MYRTLE BEACH TO VERO BEACH

The Summertime Confederate Front

The briefer said that there would be nothing significant along the route of flight; clear to 3,000 scattered, visibility eight miles, with a chance of scattered showers in Florida. There was a weak cold front across the route, but it was diffuse and not really identifiable. In all, a typical southeastern summer day.

The smooth reverie at 6,000 feet did not last long. South of Savannah, I could see what appeared to be cumulonimbus tops, and the Stormscope in the airplane was indicating a lot of electrical activity ahead. The controller said that there was a solid line of thunderstorms from twenty-five miles east of Jacksonville all the way across Florida and into the Gulf of Mexico southwest of Tallahassee. A pilot flying at 17,500 chimed in, saying that he had been trying to find a way through the line and had found it solid from Tallahassee almost to Jacksonville as he sought a path to south Florida. Another pilot said that he had come around the eastern edge, twenty-five or thirty miles out to sea, and that this had worked okay.

The controller's suggestion was for an easterly deviation. That violated a bunch of guidelines for me, though. The storms were moving to the east, so the distance from shore I'd have to go would be ever increasing and the trip around would be on the side toward which the activity was moving, where storms often demand more berth. Then there were all those sharks, and not even a floating cushion in the airplane. I felt there was no choice but to try going around to the west. The controller approved that, with a "you ain't gonna make it buddy" tone in his voice.

The Stormscope supported the decision, showing more of a western limit to the activity than the controller suggested. As I neared the activity, it even looked good to the eye. I was at 6,000 and there were some clouds that appeared to be based

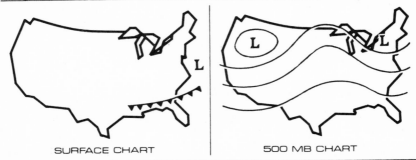

SURFACE CHART   500 MB CHART

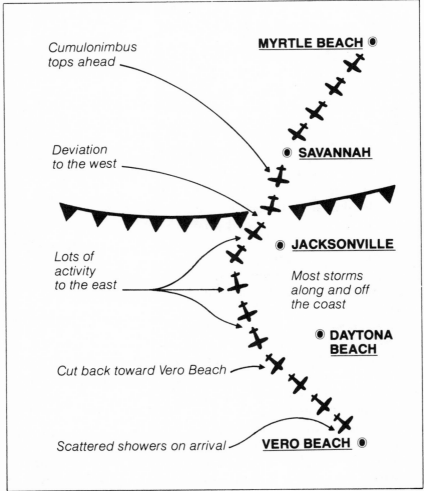

Cumulonimbus tops ahead

MYRTLE BEACH ◉

Deviation to the west

◉ SAVANNAH

Lots of activity to the east

◉ JACKSONVILLE

Most storms along and off the coast

◉ DAYTONA BEACH

Cut back toward Vero Beach

Scattered showers on arrival

VERO BEACH ◉

at about 4,000, so I requested a lower altitude for a better view; 3,000 was approved.

In contact with Jacksonville Approach Control, I found that their radar showed a good path around just to the west of that city. There was rain all around, but it was falling from higher clouds and the visibility wasn't too bad. There was heavy rain and lightning off to the left.

As I moved southward past Jacksonville, the controller assigned a southeasterly heading to go back to Victor 3, the airway along the coastline. That probably looked okay on his scope, but it didn't look at all attractive to me. The Stormscope showed a lot of activity in that direction, and the sky had a dark, ominous appearance. I requested a southerly heading instead, because that looked a lot more viable. The stuff was moving east and I wanted to stay west of it.

In retrospect, the Jax controller's radar probably extended only to a point about halfway between his airport and Daytona Beach, and there was considerable activity just to the south of his coverage area. This was verified when I switched over to Daytona Approach Control. They had a lot of activity to the north, northeast, and northwest, and there the controller's job had become one of arbitrating the differences between storms and airplanes. Some pilots were flying rather lengthy detours, others seemed of a mind to plow through. One who decided to go up the coast, through the activity, came on the frequency with his voice up an octave and requested an immediate return to Daytona. Even at a low altitude, 2,000 if I recall, this pilot had found the turbulence a bit much. My primary need was for approval to fly through a restricted area that was blocking a storm-free path, and the controller obtained this for me promptly.

The National Weather Service issued a Sigmet for these storms at 1450 EDT, about an hour after I had first started detouring around them, and probably an hour and a half after the controller had told of a solid line of activity across Florida. As I moved south of Daytona, they thinned out somewhat and I could move back close to the coast. I did feel that they were dissipating at this time. In fact, in watching them this day, and in observing a similar situation the next day, I felt as if the storms were building rapidly inland a few miles, maturing at about the time they reached the coastline, and then weakening as they moved out over the ocean. That's a logical progression, anyway. And where there might have been a more or less solid line of rain, the actual thunderstorm cells were scattered to broken.

There was some increase in shower activity near Vero Beach, and on landing there I relearned an old weather lesson that should never have been forgotten.

There were heavy showers to the south and west of the Vero Beach airport, moving toward the airport. Runway 11 was active with the wind from 160 degrees at 15 knots. I came close to asking the controller for a landing on Runway 22 but didn't because there was other traffic in the pattern. That was my mistake. Just as I flared for landing, the controller called and said the wind had shifted to 230 at 25 knots. The airplane was handling okay and the crosswind seemed within limits so I completed the landing, but it would have been much nicer on Runway 22. With thundershowers to the south and west and the surface wind from the southeast, I should have known that there would be a windshift to the southwest or west when the influence of a thundershower was felt at the airport. I should have planned the approach to accommodate that meteorological fact.

This activity was caused by a combination of a weak front, a weak trough aloft, and a pocket of cool air aloft that created just enough instability for that warm and moist Florida air to blossom into thunderstorms. The rain activity was widespread, but the occurrence of actual cells within the rain area wasn't so frequent and the storms weren't severe because not one of the contributing factors was strong. In all, a typical showery Florida afternoon.

JULY—TRENTON TO LEXINGTON

Frontal Harassment

Clear as far as Elkins, the man said, with some high clouds after that. The Lexington forecast at the ETA was for 5,000 scattered, 10,000 broken, seven miles visibility with occasional light rain lowering the visibility to three miles. He said that there was a stationary front south—Lexington would be rather

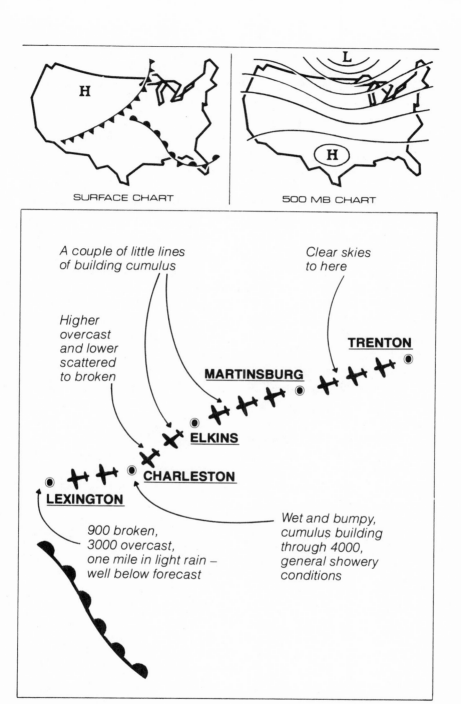

SURFACE CHART

500 MB CHART

A couple of little lines
of building cumulus

Clear skies
to here

Higher
overcast
and lower
scattered
to broken

TRENTON

MARTINSBURG

ELKINS

CHARLESTON

LEXINGTON

900 broken,
3000 overcast,
one mile in light rain –
well below forecast

Wet and bumpy,
cumulus building
through 4000,
general showery
conditions

close to it—and the winds aloft would be westerly at 15 to 20 knots.

The weather was bright and clear as far as Martinsburg, then some cumulus started developing and building through my 6,000 foot cruising level. Next came a couple of lines of little cumulus, tops to about 10,000 feet. It was a little bumpy going through, but nothing special. After the second line, I flew into an area with a high overcast and lower scattered to broken clouds.

I was watching the temperature while flying through the cumulus and was interested in a temperature variation of over 5 degrees C at 6,000 feet. In cloud it was warmer, out of cloud it was colder—a nice example of cumulus bubbling up into cooler air, but there was apparently a complete lack of upper level support to let them build into something substantial in this area.

On down the line things weren't so stable. After passing Charleston, I started passing through some more clouds, and through occasional rainshowers. In answer to a question about weather, the controller said that he wasn't painting any activity but that radar-equipped aircraft had been deviating to the south between Charleston and Lexington. By this time, some rather wet and bumpy clouds were building through 6,000 feet, and on the strength of a 4,000 broken, 12,000 overcast report from Lexington, I got clearance to descend to 4,000. The thought was that four would be below the cloud bases, but that wasn't the case at all. Conditions at that altitude were just as they had been at 6,000—in and out of cumulus, wet and occasionally a little bumpy. The controller was not painting any weather but the Stormscope in the airplane did show some electrical activity ahead. It appeared to be rather close on the scope, and at one point I deviated to the north based on what I saw on the Stormscope and a more friendly appearance of the sky in that direction. It didn't make much difference in the ride, and what the Stormscope was showing as close was a very strong storm actually quite some distance away. The device will do this.

The weather that I got from the Lexington automatic terminal information service was interesting: 900 broken, 3,000 overcast, visibility one mile in light rain and fog. The wind was light and southerly. That was quite different from the last forecast. I took this as an indication that the front was farther north than had been anticipated.

Before reaching Lexington, I flew back into a situation with lower clouds plus a high overcast. The visibility at 4,000 was quite good except in areas of rain. It was clear to the eye that there wouldn't be any thunderstorm problem getting into Lexington, and the weather there was as reported on the ATIS.

The rainshower area to the east of Lexington had some of the appearances of a front, but there was no front depicted there. It could easily have been a rather weak wave on a stationary front, which is always something to look for, but the official National Weather Service map for this day didn't show the front that the briefer had told me about on the phone. Instead, it showed a warm front to the west. In that case, the rainshowers and below forecast weather at Lexington could be charged to instability on the slope of that warm front. We'll continue the discussion of this day's weather on the following flight.

JULY—LEXINGTON TO LAKEVIEW

Activity on the Front

This flight, from Lexington to Lakeview, Arkansas, was a continuation of the previous trip, and when I called for a briefing I had some idea of what to expect. The front was said to be southwest of Lexington, and it seemed to be doing things farther to the north than they had originally thought. It was now drawn on the weather map as a warm front, and a lot of stations down the way were reporting thunderstorms. The briefer did not have any radar information, but he did have a Sigmet for the area that advertised thunderstorm tops to 55,000 feet. The briefer strongly suggested a southerly route, by way of Memphis, instead of a direct route over Paducah, Kentucky. His theory was that there were more stations reporting storms along the direct route. I bought his theory and filed the flight plan by way of Memphis.

Once airborne and above the low level grunge that was restricting surface visibility, the view strongly suggested that I not fly the route suggested by the man at the FSS. It was black

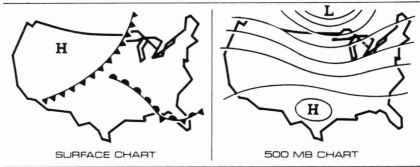

SURFACE CHART | 500 MB CHART

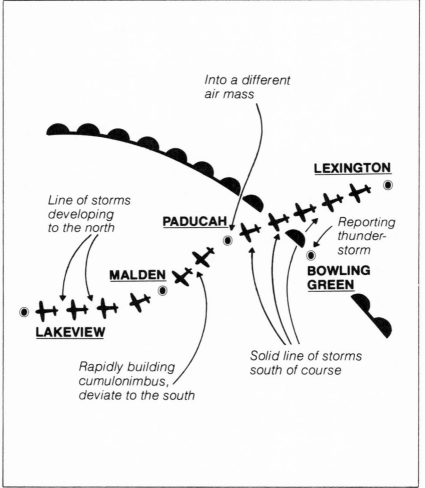

Into a different air mass

LEXINGTON

Line of storms developing to the north

PADUCAH

Reporting thunder-storm

MALDEN

BOWLING GREEN

LAKEVIEW

Rapidly building cumulonimbus, deviate to the south

Solid line of storms south of course

to the south and southwest and the Stormscope was depicting intense electrical activity. The way to go was west, at least for a while; the controller agreed and provided a clearance to go in that direction.

The storms to the south were obviously quite strong, and at one point the controller said something about tops to 64,000 feet. They were clearly worth staying away from, and when the controller asked if I could go over and fly the airway that went over Paducah, I agreed because it looked like the best way to go. There appeared to be some rain in that direction, but it was falling from much higher clouds and the airway looked okay.

When about 70 miles west of Lexington, I encountered the strongest subsidence I've ever seen well away from a thunderstorm. I had a 30 knot loss in indicated airspeed for a protracted length of time, and it certainly wasn't in an area where you could envision a downdraft from a dissipating storm. It had to come from air beneath being sucked into that monstrous storm, and some subsidence was to stay with me for almost 200 miles. The indicated airspeed didn't get back to normal until I was on the south side of the activity.

The official NWS map issued for the day depicted a warm front in a northwest-southeast line over eastern Missouri and Arkansas. I was flying north of an east-west line of activity from Lexington to Paducah, and some of the storms were very strong, but in many areas the rain appeared to be falling from a rather high overcast. Even where the overcast appeared high, there was a lot of cloud-to-ground lightning. It had more the appearance of an occlusion, where the frontal slope was high and nothing was moving very much. And, again, it had more an east-west orientation.

In trying to put together a better mental weather map, I started checking surface winds and comparing altimeter settings. On the latter, I found that the pressure had dropped slowly to Lexington and then had started rising slowly. The surface wind at Bowling Green, south of my course, was from the west. At Evansville, north of my course, the wind was from the south. That didn't relate to any imagined situation, and I decided that the surface winds were influenced more by the thunderstorm activity than by general weather patterns. As far as pressure went, the changes were so minor that they didn't have much significance. The general appearance of the weather remained the best clue to what was going on.

Near Paducah, it appeared that I was coming to the end of the dark gray line to the south. It also appeared that I'd be passing through some sort of line there. Paducah was reporting a thunderstorm, but with high ceilings, and it looked bright enough ahead to continue without a qualm.

There was only light rain from a high overcast in the vicinity of Paducah, and where the visibility had been good to the north of the line of weather, it turned hazy as I passed to the south. It was obviously a different air mass and I thought that there would be no further thunderstorm encounters. I was wrong.

I could soon see some activity building to the southwest of Paducah. It appeared to be building very rapidly, and was already through 18,000 or 20,000 feet. I asked the Paducah approach controller for a deviation to the south. He denied the request because there was no radar contact, and only after some rather forceful discussion did he do the necessary work to approve the deviation. I had no intention of flying through the blossoming cumulus, and would have had to switch to VFR and go bump along beneath all clouds if the request had been denied.

This next batch of weather was entirely different from what I had flown north of in Kentucky. There was building from down low—cumulus bases were probably 2,000 to 3,000 feet —and it was just beginning. A line was actually forming, and after picking my way through some rather low spots, flying at 6,000 feet, I found myself on the south side of what appeared to be an east-west line of rapidly developing thunderstorms. I was talking with the Center controller by the time I had passed Malden, and there was a lot of chatter about the line of storms. Aircraft at high altitudes were even beginning to deviate around the weather, and it was clear that the afternoon was going to be a thundery one in that part of the country.

The difference between this batch of storms and the first batch was in age. The triggering mechanism in Kentucky was a warm front that had been there for a while, and the system was an old one with a lot of rain and some strong cells mixed in. The activity in southern Missouri and northern Arkansas was new and building, sparked by a cold front. It was easy to see, and had the classic appearance of strong summertime activity.

Once south of the line, the trip on to Lakeview was uneventful. Later that evening, though, strong thunderstorms rumbled through the area.

The weather map didn't depict a low at the junction of the cold and warm fronts but there was a weak trough aloft north of the area and some southerly movement of cold air aloft. There must have been some large scale ascent ahead of the systems depicted on the map, and the warm summertime temperatures at the surface and the cool air aloft completed the picture. This was quite a good example of how diffuse systems can be in the summertime, when the lines drawn on the map often fail to translate to the clear-cut definition that we see at other times of the year. Having experienced the situation this day, I'd have depicted an occlusion from Paducah over to south of Lexington, curving back northward just east of Lexington. And I'd have drawn it as a cold front across northern Arkansas where the storms were building rapidly. At least that's what it looked like from an airplane, which is a better observation post than an office.

JULY—WORCESTER TO TRENTON

Weak System

This flight was influenced to a degree by the weather guesser's product of the previous day. The weather was perfect for a VFR flight to the Worcester area, but I had to choose between landing at the large airport, or at a small one with no IFR services and enough high terrain around to preclude an instrument departure the next morning if conditions should be very low. The forecasters were predicting that a low would move into the area and be just southwest of Worcester during the night, which would make my departure from the small field questionable the next morning. I landed at the large airport and went about my business.

I felt a touch foolish when the next morning dawned bright and clear. Cool, hardly a cloud, and a brisk north wind gave the air more the appearance of fall than summer.

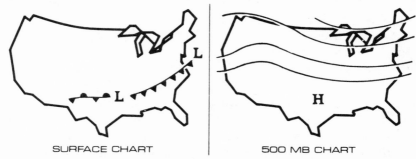

SURFACE CHART   500 MB CHART

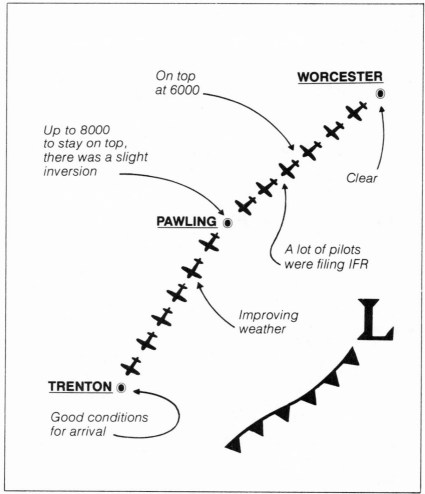

On top
at 6000

WORCESTER

Up to 8000
to stay on top,
there was a slight
inversion

Clear

PAWLING

A lot of pilots
were filing IFR

Improving
weather

L

TRENTON

Good conditions
for arrival

The FSS briefer read forecasts, and it was apparent that at least the Worcester forecast should have been in the wastebasket instead of on a clipboard for reading. It called for good ceilings but restricted visibilities in fog—two miles until nine, then four miles. From my vantage point, you could see forever. I could only feel that a front had passed Worcester, or a low had moved to the east, contrary to what the forecaster expected. En route, the weather was okay—high broken with five to seven miles visibility—and Trenton, the destination, had a forecast of 3,500 broken with five miles visibility, chance of rainshowers, wind shifting from southwesterly to northeasterly at about the ETA.

An IFR flight plan was filed, mainly in anticipation of anything that could develop along a front. Flying at 6,000, I wasn't long in coming up over some clouds, which wouldn't have been expected from all the reports and forecasts of high broken, but which were quite logical in a frontal zone. Bases looked to be about 1,500 to 2,000 feet above the ground, and the air traffic controllers were handling a lot of impromptu IFR flight plans from pilots who expected better conditions than they found.

The tops built through 6,000 feet about 60 miles west of Worcester, and provided some bumps in the cumulus clouds. There was no precipitation around, though, and the ride was characterized by some upward action in the cumulus, some subsidence between them, and mild turbulence. The cloud layer below finally became solid, and I moved on up to 8,000 feet to stay in smooth air. There was a slight inversion between 6,000 and 8,000 and the ride at the higher altitude was serene and smooth. The farther southwest, the better the weather, and by the time I reached Trenton it was quite good. The wind there was out of the north.

The low was just south of Long Island, with an associated cold front, and there was a weak trough aloft, around a low far to the north. It wasn't actually a front that we flew through southwest of Worcester, but rather a collection of clouds that developed behind the front, probably aided by the high terrain just to the north of our route of flight. The inversion prevented the clouds from growing much past 8,000 feet and from becoming more than a minor factor in the flight.

AUGUST—TRENTON TO SAGINAW

Trough

There was a weak low off the Jersey coast with a stationary front extending to the southwest. The forecast was for marginal VFR conditions in the area for a 2 P.M. departure, with continually improving weather for the westbound journey. The forecasts past Buffalo were for unlimited ceilings and visibilities. Winds aloft were to be relatively light, from the southwest or west.

Takeoff was with 500 overcast and a mile and a half visibility instead of the forecast marginal VFR conditions, and cloud tops were at 2,500 feet. There was a substantial amount of haze above the lower clouds, with in-flight visibility probably not over four miles. A high broken layer completed the picture. It was typical of summer weather to the northwest of stationary fronts along the east coast.

The cloud tops increased in height to the west, and over Williamsport at 8,000 I was flying above a layer that probably topped at 7,000. The +12 C temperature at 8,000 compared with +25 C on the ground for pretty stable air. In verification of this, the tops were relatively flat.

The temperature at 8,000 dropped to +7 C near Buffalo where the surface temperature was up to +27 C, and there was quite a collection of cumulus in the area in honor of the increasing instability. Some deviations were necessary to avoid cumulus that had built above 15,000, to 18,000 feet, and the sky became rather congested before I flew out into basically clear sky just to the west of Buffalo. There had been a little line of building stuff there, about thirty or forty miles through, and I thought that it could develop into something later in the afternoon.

In checking weather with the Buffalo FSS, I found that there was some thundershower activity around London, Ontario, and the London surface weather observation verified this with the remark: "dark northwest through northeast."

The tops of a couple of cumulonimbus were visible from quite a distance east of London, and it looked as if they would be just north of our course. The controller said first that no

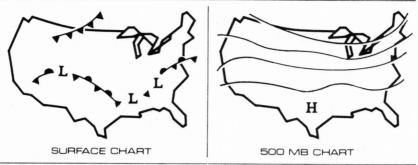

SURFACE CHART | 500 MB CHART

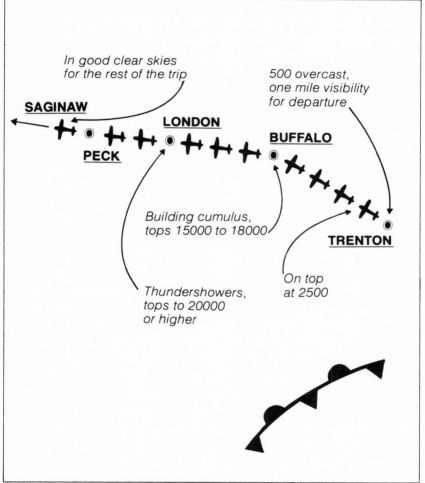

In good clear skies for the rest of the trip

500 overcast, one mile visibility for departure

SAGINAW

PECK

LONDON

BUFFALO

Building cumulus, tops 15000 to 18000

TRENTON

On top at 2500

Thundershowers, tops to 20000 or higher

aircraft had been deviating, and he later called and said that some aircraft were now deviating around weather in the London area. He added that he was painting very little precipitation, but also that his radar was set to eliminate most precipitation return.

Closer to London, it was apparent that we'd need to go slightly south, and this was approved by the controller. The tops of the build-ups appeared to be above 20,000 by this time, and there was quite a bit of rain falling but only occasional lightning. The cumulus were just barely making it through the freezing level. The relatively light southwesterly winds weren't providing a lot of moisture, and the lifting action was weak. The only positive factor for development was instability in the lower levels.

This area of showers ended abruptly near Peck, and the flight on to Saginaw was in excellent weather.

While flying along, I thought to connect the activity near Buffalo with that near London for some sort of east-west trough. In the FSS at Saginaw, the latest surface chart did show a trough line, but it was northeast-southwest, near London. That explained the activity that was trying to develop in that area; perhaps the building cumulus over around Buffalo were related to some effect of Lake Erie, just west of Buffalo, although this usually isn't much of a factor in the summertime. The 500 mb chart for the morning did show a small trough aloft over central Ohio, and the activity around Buffalo later in the day could well have been associated with this.

AUGUST—GREEN BAY TO NEWBURGH

Sluggish Cold Front

The cold front had moved through Wisconsin on Wednesday morning so, when the briefer said that it was at London, Ontario, on Thursday morning, I knew it was a slow-mover. That 13 knot average just didn't qualify it as a fast-mover.

There had been some thunderstorms with the front on Wednesday, and on Thursday morning the radar summary depicted a broken area from east of Detroit over to the Buffalo area, no lines, maximum tops to 33,000 feet. The surface

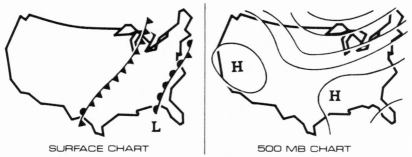

SURFACE CHART | 500 MB CHART

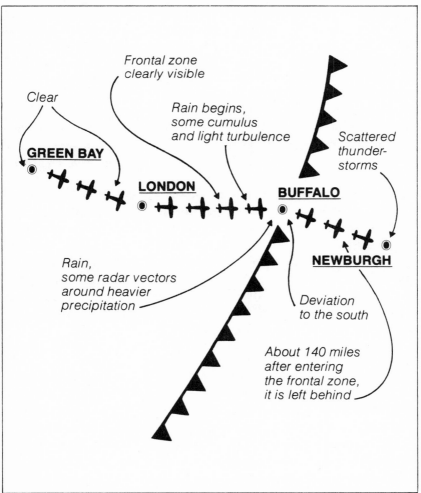

Frontal zone
clearly visible

Clear

Rain begins,
some cumulus
and light turbulence

Scattered
thunder-
storms

GREEN BAY

LONDON

BUFFALO

NEWBURGH

Rain,
some radar vectors
around heavier
precipitation

Deviation
to the south

About 140 miles
after entering
the frontal zone,
it is left behind

weather was good as far as London where it became cruddy and rainy. This extended over to just east of Buffalo. The east coast was under the influence of a stationary front that had been around for several days, and the forecast for arrival was 1,000 broken, 2,000 overcast, four miles visibility, occasionally two miles in light rain. The chance of a thunderstorm and lower conditions was suggested after 2 P.M., which was my ETA at Stewart Field, Newburgh, N. Y.

To begin with, winds were to be from the west at 25 knots at 12,000 feet. Later in the flight and at lower altitudes they would shift to southwesterly and then to southerly on the east side of the front. The initial concern was with the 12,000 winds, as I'd be flying across Lake Michigan, so I went to 11,000 feet. That would minimize the exposure to a splash in case of mechanical trouble over that 40-nautical-mile-wide pond.

The weather to begin was beautiful. I could easily see across the lake, and many ships were plying back and forth. The groundspeed on the first leg was nice, too, at 172 knots. That meant the wind was 35 knots, and very much appreciated.

On the east side of Lake Michigan I flew up over a lower deck, but this was just morning stratus. Its presence at 10 A.M. did suggest a rather lazy northwesterly surface flow behind the front, and a bit farther on, after the undercast was gone, the surface flow according to factory smoke looked to be northerly or even slightly to the east of north.

En route weather checking revealed that the front had moved well east of London, and a later radar report than I had received at Green Bay showed scattered rain and rainshowers in central New York and Pennsylvania. There were no thunderstorms mentioned in this report.

The rear end of the frontal zone was clearly visible from seventy miles west of Buffalo, and when fifty miles west some cumulus clouds were visible building from below into a higher overcast, based at an estimated 13,000 to 15,000 feet. I didn't want to fly through the front at 11,000 so I left that level for 7,000 as the rain area was approached, thirty miles west of Buffalo. Among other things, I wanted the better weather advisories that are usually available from approach control radar controllers, and at 11,000 I would have been with the air route traffic control center controllers all the way through the front.

The Buffalo controller said that there was a lot of weather around and that some pilots were flying through while others were deviating. There were numerous pilot reports of heavy rain, but none of turbulence. The Stormscope in the airplane was inoperative on this trip, so I had no on-board information source other than the ADF. It had static on the low band, an indication that there were some thunderstorms somewhere. There were no Sigmets.

I deviated around some build-ups that were visible, but the airplane was in cloud most of the time so visual information wasn't anything to depend on. To the east of Buffalo, the controller said that there was weather ahead and that Rochester Approach Control would have better information on it. Indeed they did, and after hearing word of an area of showers and then a couple of lines of weather, I wondered for a moment about the situation. On the negative side were occasional pops in the VHF radio, indicating electrical activity nearby, and then there was that word about weather ahead. On the positive side was the fact that the frontal zone was quite broad, and that the airplane had already moved through a lot of rain without encountering anything more than light turbulence. Also, there was no indication of anything severe from the NWS, and the information that I had gleaned before and during the flight suggested that the weather wouldn't be too turbulent.

When asked for a vector around anything that looked heavy, the Rochester controller told me to steer 180 degrees for a while and then cut back to the left. This was done, and it apparently kept me clear of what he had described as a line of weather. That left one line to go and, in answer to further questioning, it was described as light. I had moved on down to 5,000 feet by this time, was occasionally clear of clouds, and could see that the situation didn't appear too bad. Everything appeared to be light gray in the area.

Rochester handed control of the flight off to Elmira and the controller there said that all the weather was behind, steer 100 to intercept the airway. Done, the heading took the airplane through the darkest and bumpiest cloud of the day. There was very little rain there, so the controller didn't show it on radar, but it was the leading edge of the area of frontal weather, on the east side of the area of lifting and instability, and it shook the airplane around a bit.

East of that, the weather was clearly of a warm sector: blue sky, lots of building cumulus, and some thunderstorms off to the east. The altimeter setting started back up, too. The relatively small pressure drop, from 30.07 in Wisconsin down to 29.99 at Buffalo, ended and it moved back up to 30.07 on the east coast —another indication that the frontal system was not strong.

The thunderstorms to the east were more organized than they looked from a distance, and were actually in connection with the stationary front that was finally moving off the coast. They almost formed a line and moved from west to east across the New York area before I got to Newburgh.

Looking at the weather charts for the day, the cold front was almost to Buffalo at 8 A.M. EDT, a bit after I checked weather, with a weak trough aloft. The temperature at the 500 mb level was −10 C which, given maximum surface temperatures of about +27 C, did reflect some instability. But there was nothing there for really big vertical development, and apparently none occurred. It was interesting that the "Today" program weatherman offered a rain probability chart that morning showing the likelihood of from one to two inches of rain over central New York and Pennsylvania.

AUGUST—TRENTON TO WASHINGTON TO TRENTON

Afternoon Thunder

A stationary front had been languishing south of the area for days, weeks really, and the weather had been hot and humid with thundershowers and thunderstorms. There had been a lot of action at night, diminishing in the morning with redevelopment by evening. The early morning briefing the day of the flight called for a continuation of this trend. Thunderstorms were mentioned as being in the area of Wilmington, Delaware, and southern New Jersey. The forecasts for all stations were about the same: There would be low clouds and restricted visibility in the morning, followed by some improvement, with the chance of thunderstorms forecast throughout the day. The

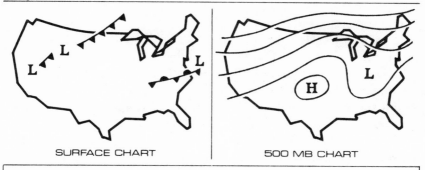

SURFACE CHART | 500 MB CHART

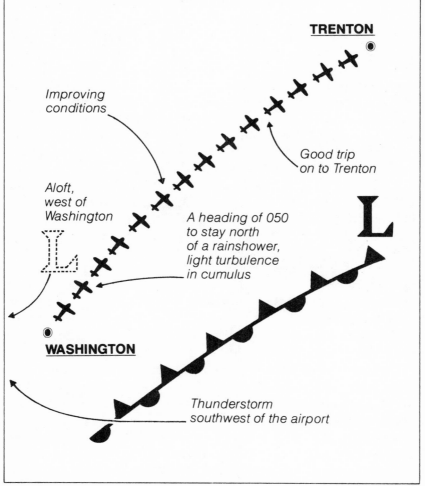

TRENTON

Improving
conditions

Good trip
on to Trenton

Aloft,
west of
Washington

A heading of 050
to stay north
of a rainshower,
light turbulence
in cumulus

WASHINGTON

Thunderstorm
southwest of the airport

briefer at North Philadelphia had nothing more than an old radar summary chart to use in the briefing, so before the 8:15 A.M. takeoff I called the Washington FSS to get a reading from their radar. They showed no activity to the northeast.

The ceiling at Trenton was about 300 feet and tops were 2,500 feet. It was almost clear above—only some high scattered —but the haze layer extended upward to probably 10,000 feet. The situation looked very flat—not a building cumulus anywhere. Washington National was reporting 700 scattered, 2,000 broken, and three miles for our arrival at 9:25 A.M.

Washington business finished, the view of the sky from downtown at 12:40 P.M. strongly suggested that things had changed. It was dark to the southwest, and building cumulus could be seen in all quadrants. Rain was falling when I reached the airport, and the FSS confirmed that there were a lot of scattered showers around the area, especially to the southeast. I was headed northeast, so that didn't sound too bad. At the airplane, a thundershower was both visible and audible just to the southwest of the airport, and it appeared dark to the east.

The initial clearance was to depart on Runway 3, turn right to a heading of 050, and maintain 2,000 feet. It looked okay in that direction and, in fact, even the storm to the southwest had a rather lackadaisical appearance. Appearances deceive, but it sure didn't look like anything more than a thundershower, and there was no breeze from it across the airport even though it was just to the southwest.

Once off, I could hear other aircraft requesting deviations around weather, and when the controller called with word to fly a heading of 090, I told him that it didn't look good in that direction and that I'd like to fly 050 for a while. The heading was approved, and a clearance to climb to 5,000 feet meant that I'd be in cloud instead of below the bases, as was the case at 2,000 feet. The airplane passed through a nice fat cumulus with some rain during the climb to 5,000, and the updraft in the cloud added about 700 feet a minute to the normal rate of climb for about one minute. Turbulence was very light.

Not far northeast of Washington, the situation improved markedly. Where there had been a good collection of cumulus with tops probably nearing 20,000 and some already to the cumulonimbus stage, the weather forty miles northeast of

Washington was typified more by rather slowly building cumulus, higher broken, and quite a bit of haze. The situation remained like that until near Philadelphia, when the build-ups started squirting upward at a good rate again. In flying through the smaller ones, though, I didn't find much turbulence. Trenton was reporting 3,000 broken and three miles in haze for landing there.

The surface temperature was +30 C over much of the area and the temperature at 6,000 was +16 C so there was instability.

The map for the morning showed a low to the south of Long Island with a low aloft inland, over West Virginia. Neither pattern was strong. The winds aloft were light and variable through 6,000 feet, so the low level moisture was coming more from the ground than from any circulation. The ground had plenty to offer, too, because of the abundance of rainfall during the preceding days.

The low at the surface, combined with the one aloft, didn't offer quite enough action to trigger storms during the morning, but when the sun's heating had been added to the equation for a while there was enough to make the clouds grow rapidly into cumulonimbus. (After I left Washington, it rained so hard that some streets flooded and the roof of at least one government building leaked.) The storms were not severe, but there would surely have been a good beating in one for a light airplane. With such heavy rainfall rates, though, they would have shown pretty clearly on the controller's radar scope.

The position of the low aloft explains why there was more of an outbreak of activity near Washington than in the Philadelphia area. The movement of air up and around that low contributed more to lifting in the Washington area than farther to the northeast.

AUGUST—TRENTON TO COLUMBUS

Cold Front That Didn't

The weather briefing for this late afternoon flight started in my New York office. A colleague had come in from the west the day before in his Bonanza and reported that a cold front out

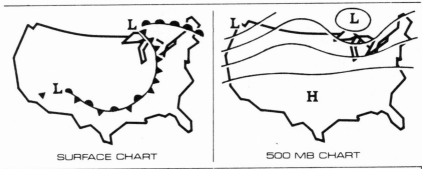

SURFACE CHART   500 MB CHART

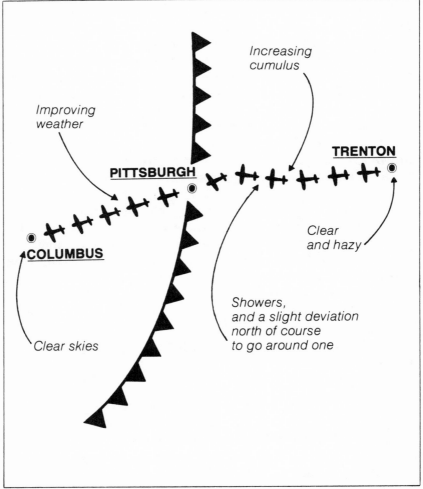

Increasing
cumulus

Improving
weather

TRENTON

PITTSBURGH

Clear
and hazy

COLUMBUS

Showers,
and a slight deviation
north of course
to go around one

Clear skies

that way was pretty active. "Got my plow cleaned," he said. "Stuff hit the ceiling, and paint was knocked off the prop spinner in the heavy rain." Regardless of how stoic one might be about the elements, such a description of a front tends to wave a flag.

The lady at the flight service station waved more than one flag when I called for information. In a most somber voice, she told of a line of thunderstorms approaching western Pennsylvania, a severe thunderstorm watch for Ohio and western Pennsylvania, and the possibility of air traffic control instituting a severe weather avoidance plan until 1900Z. (In such a plan, they route traffic around general areas of severe weather.) In answer to my question about activity on the radar summary chart, she said that there were only some widely scattered build-ups in eastern Ohio and western Pennsylvania, but, regaining her strength, she said that these could build "explosively." The weather at all stations was reported as good VFR, and forecasts were for it to remain so except in occasional thunderstorms. The front was actually west of Columbus, and the winds aloft were southwesterly at low altitude shifting to westerly at 9,000 feet.

Despite the ominous talk of severe thunderstorms, it seemed okay to launch. There was nothing out there at the time, the front didn't seem strong (the southerly surface winds were only 10 to 15 knots ahead of the front, and the westerlies behind it were light), and what I could learn about the upper level patterns from the winds aloft didn't suggest a lot of trouble. Also, I would have the distance between Trenton to Pittsburgh to get more information and study the situation.

The weather was clear but hazy at the airport, and at the assigned IFR cruising level of 6,000 feet it was smooth, sunny, and hazy with an occasional cumulus cloud. The surface temperature had been +30 C; it was +19 C at 6,000 feet so if the air was unstable it wasn't overly so.

Around Johnstown, the collection of cumulus increased, prompting a question about weather to the controller. He said that there had been some build-ups a while back but they were gone now. Some airplanes were still deviating in the area, but nobody was reporting anything significant.

Closer to Pittsburgh, the cumulus seemed to be gathering

together and building to a higher level. I moved through a couple with only light rain and light turbulence, and then came into an area where there were much darker clouds ahead. The controller said he was beginning to show something at 12 o'clock and five miles, and when I asked if a 30 degree deviation to the right would avoid that, he said that it would.

The general cloud deck in the Pittsburgh area was reported at 4,500 feet, so a request for 4,000 seemed in order. It was granted, and at four I was below the cloud base, flying in a rather gloomy sky with some showers around. There was no turbulence, and west of Pittsburgh the cloud cover started diminishing. The headwind increased slightly, too, indicating a shift to a more westerly wind. Some cumulus build-ups, probably not over 10,000, were around, and when flying through a couple I noted that there was very little moisture in them. They hardly wet the leading edge of the wing as the airplane passed through, and perhaps a general lack of moisture kept the weather from developing as expected this day. Columbus was clear.

The early morning weather map for this day positioned a surface low over Lake Superior, with a low at the 500 mb level just above the surface low. There was a squall line in western Ohio, extending down into northwest Kentucky. These patterns, if they had continued, would indeed have created thunderstorm activity throughout the day.

The picture changed substantially by late afternoon, though. Both lows, the one at the surface and the one aloft, tracked northward instead of eastward. The cold front remained an identifiable entity on the chart, but it weakened and slowed as the low moved to the north. Despite all the rumbling about the possibility of severe weather in the afternoon, there was nothing there to generate anything more than the showers that were encountered in the Pittsburgh area. A weak cold front, a low far away, a limited supply of moisture, and relatively stable air just don't combine to create a lot of action.

AUGUST—TRENTON TO SKY BRYCE

A Grungy Old Day

The plan was to fly from Trenton to Dulles airport near Washington to pick up a passenger, and then go on to Sky Bryce, Virginia, where there's a nice little 2,400 foot strip snuggled up in a relatively narrow valley. That last part had to be VFR—no instrument approaches in those hills—and the briefer said that it should work okay. There was a stationary front to the south, and while all stations were reporting IFR conditions at 9 A.M., the forecast was for 3,000 broken, 8,000 broken, and four miles visibility with a chance of thunderstorms by noon.

At the Trenton airport, the reported weather was 600 overcast and one mile, but the fog had burned away and it was really just hazy with a couple of miles visibility. The thought was that the clouds would burn off and the day wouldn't be bad at all.

The picture changed after departure. At 4,000 feet I was in cloud most of the time from north of Philadelphia to Dulles, and the clouds became progressively bumpier as the airplane moved southwestward.

Dulles was very far from its forecast, with 800 overcast and one mile visibility. The sky didn't have the look of something that would improve soon, and having just landed I knew that the cloud deck was something more than a collection of early morning fog and stratus. The weather over in the Shenandoah valley, at the closest reporting station to Sky Bryce, was pretty good—3,000 broken and five miles—but the FSS person in Washington said that showers and thundershowers were breaking out along the ridges to the west. The activity did seem more to the north than the west, and the best procedure seemed an IFR flight to Luray, Virginia, only about thirty miles from Sky Bryce. I could shoot an instrument approach there and go on to Sky Bryce VFR if possible. If not, I'd land at Luray and either wait for improvement or start hitchhiking. The tops were about 5,500, building rapidly, and at 6,000 I was skimming along with occasional ground contact. There were build-ups visible ahead and the controller agreed that a deviation to the north would be a good idea. The Stormscope in the air-

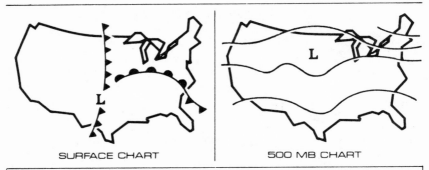

SURFACE CHART | 500 MB CHART

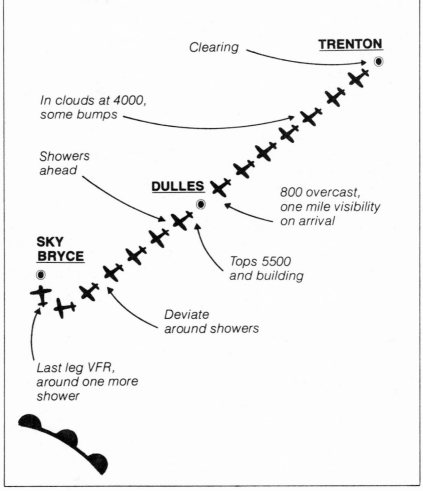

Clearing — TRENTON

In clouds at 4000, some bumps

Showers ahead

DULLES

800 overcast, one mile visibility on arrival

SKY BRYCE

Tops 5500 and building

Deviate around showers

Last leg VFR, around one more shower

plane was not displaying any activity, so it was apparently still in the shower stage. Once I zigged instead of zagged and flew through a shower; it wasn't too bumpy, and the rain was only moderate.

The approach to VFR conditions at Luray was successful, and at 3,000 feet I could see the tops of the ridges that had to be crossed going on to Sky Bryce. The controller said that there was a cell right over the airport moving eastward, but there were several alternate plans available if the VFR flight didn't work. A continuation was in order. A detour to the south took care of the cell the controller mentioned and some judicious highway following and landmark identification finally got the airplane to Sky Bryce.

Three other airplanes were following behind, en route to the same meeting, and the weather looked like it was going to work out for them. The big ridges to the west were visible before I landed and the showers moved on to the east. That decent weather was short lived, though. Some more showers moved through, and the visibility dropped in ground fog by midafternoon and remained cruddy for the rest of the day.

This trip was influenced by a stationary front that turned into a warm front and moved to the north. In fact, where the briefer had called it a stationary front, the weather map properly depicted it as a warm front a couple of hundred miles southwest of Sky Bryce early on the morning of the flight. There was very little upper level support for the thundershowers so they remained scattered and didn't become strong.

There was an interesting weather situation two days later, when leaving Sky Bryce. The weather the day after arrival was foggy in the morning with the air tending to dry out some during the day. The visibility wasn't bad and there were fewer showers than the day before. The warm front had apparently moved off to the northeast.

The weather was checked for the following day, and the briefer said that there would be morning fog, slow improvement, and a cold front approaching the area by afternoon. Some thunderstorms were expected with the front.

A couple of people needed a ride to Dulles, and the forecast of fog the next morning caused us to make plans for automobile transport. Sky Bryce is no place to leave on a foggy morning. Still, I left a call and thought I'd look at the predawn situation before sending them away in a car.

The stars were out and the briefer gave relatively good reports for all stations at 6 A.M. but he still cautioned about the cold front that would be in during the afternoon, with possible thunderstorms. I took off at 7 A.M. for the 60 nm run to Dulles, and immediately after takeoff it was apparent that the cold front they were expecting in the afternoon had already gone through and was to the east. The air was crystal clear—you could see forever—and a fairly strong northwesterly flow was active in the area.

Their miss on the conditions in the warm front, and the total (and really inexcusable) wrong call on the cold front two days later is just another reason why pilots have to fly with complete suspicion of forecasts. In one case, the weather was worse than forecast, in the other it was much better.

SEPTEMBER—TRENTON TO TERRE HAUTE

Insignificant Front

On the morning of this flight, I didn't anticipate meteorological surprises. The weather was said to be virtually clear all the way, but weather doesn't have to be significant to be interesting, educational, or a factor in flight operations.

Flying on an IFR flight plan at 6,000 feet, I found the wind aloft to be stronger than forecast. The sun was up, and I could see that all the optimistic forecasts of good weather were true. I could also see from smoke stacks that the surface wind was light and easterly. The obvious course of action was to cancel the IFR and enjoy a low level flight across the Pennsylvania mountains on this perfect morning.

The briefer had described a northwest to southeast front across western Pennsylvania, with no accompanying weather. It was basically the remnants of a cold front that had pushed to the south from Canada a couple of days before. He called it a cold front; the official weather map for the day drew it as a warm front. My early morning weather check had suggested only restricted visibility to the west of the front, improving after about nine in the morning, and in checking weather once

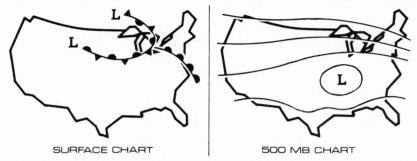

SURFACE CHART   500 MB CHART

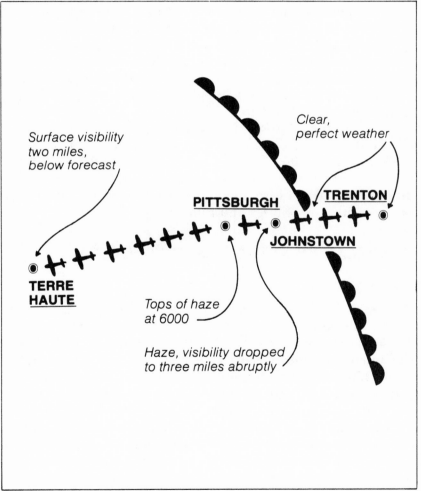

Surface visibility
two miles,
below forecast

Clear,
perfect weather

PITTSBURGH   **TRENTON**

JOHNSTOWN

**TERRE
HAUTE**

Tops of haze
at 6000

Haze, visibility dropped
to three miles abruptly

aloft I found Johnstown, Pa., to have clear conditions with fifteen miles visibility; Pittsburgh, not far west of Johnstown, had partial obscurement with three miles. All the stations in Ohio had one or two miles. No doubt, there was some change from the sparkling conditions over the mountains.

In the vicinity of Johnstown, I could see what appeared to be smoke ahead—much like you see streaming from a large area of forest fires. There were no clouds—just a band of smoky looking sky. I flew into this without a bump, and the visibility went from unrestricted to a bare three miles almost instantly. The temperature increased 8 degrees C—enough for the colder fuel in the tanks to make condensation on the bottom of the wings. There was indeed a front there.

I had never before seen a front so clearly defined by something other than clouds. It is a little misleading to say that the weather west of the front wasn't significant, too. While no clouds developed as I flew on to Terre Haute, the surface visibility continued to vary between one and two miles. Conditions did not improve later in the day, as forecast. The top of the haze was at 6,000 feet but I had to get an IFR clearance to land at Terre Haute, and traffic delays could have caused as much of a problem with minimum fuel reserves on this day as on one with low ceilings and rain.

SEPTEMBER—ST. LOUIS TO TRENTON

Warm Front

A very loud thunderstorm rumbled through St. Louis at 4 A.M., and when I peered out the window of the hotel at six it was apparent that several days of hot and muggy were giving way to cooler and drier air. That was nice for the folks in St. Louis, but I would be headed east at 10 A.M. and it was quite logical that I'd have to negotiate with the weather system along the way.

The weather was good at all reporting stations when I checked for the flight, but about 100 miles to the east of St. Louis was an area of rainshowers and thundershowers, with 60 percent coverage, decreasing in intensity. The latter remark was encouraging.

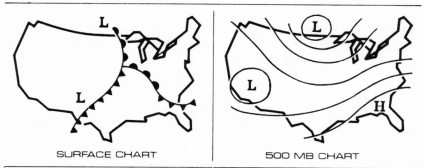

SURFACE CHART   500 MB CHART

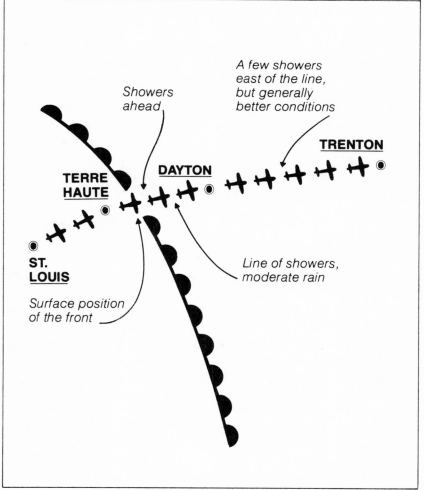

A few showers east of the line, but generally better conditions

Showers ahead

TRENTON

DAYTON

TERRE HAUTE

ST. LOUIS

Line of showers, moderate rain

Surface position of the front

Up at 9,000 feet, conditions appeared good as I progressed eastward. The air was smooth and a tailwind was moving the airplane along at 155 knots. The briefer had insisted that the storms were unrelated to any front and that a cold front would be along later in the afternoon. It was apparently a warm front, though, as had been portrayed on the "Today" program weather map. The surface wind ahead of the storm area was southeasterly; after it passed, the wind shifted to southwesterly. As I flew eastward, the pressure dropped with each report up to Terre Haute, and then it started back up. That was the surface position of the warm front, so any action would logically be east of there.

A Sigmet was broadcast that advertised scattered embedded thunderstorms, but the area of coverage was just north of my route of flight. In discussions with the traffic controller and with the pilot of a radar-equipped Twin Bonanza just ahead of me on the airway, I found that there was no significant weather on my route of flight. The Stormscope agreed, but some building and fat cumulus were visible in a line of rain that I was approaching, in the vicinity of Dayton, Ohio. All pilots flying through reported only moderate rain, but I tightened the belt and slowed the airplane to maneuvering speed anyway.

There was a period of enough rain and updraft to require a substantial reduction in power to maintain maneuvering speed and altitude, but the turbulence was very light. The line of rain was only fifteen or twenty miles through, but there were building showers to the east of there, and several deviations were required to get around the more enthusiastic build-ups. It was just after noon when I passed Dayton, and I thought it good that I had come through early in the day. The activity had been strong when it passed through St. Louis very early in the morning, it had weakened, and it might well increase in strength again by late afternoon. The temperature at 9,000 was +16 C, reflective of less than a 2 degree C per 1,000 foot drop at low altitude over the area, so nothing big was to be expected. But still it could have done enough to bounce a little airplane around a lot.

The build-ups through 9,000 continued on to Pittsburgh where the situation became one of placid, layered clouds. The temperature aloft decreased as we flew along, as did the surface temperature.

This flight gave an excellent example of wind aloft increasing in velocity in the frontal zone, and then shifting to a southerly direction on the north or east side of a front. The ground speed started out at 155 knots, with little drift on an easterly heading. It increased to 166 knots for a 72 nm leg that included penetration of the frontal slope, and then for the rest of the trip the groundspeed settled back to 145 knots with about 10 degrees of drift to the north.

The National Weather Service map for the day depicted the warm front clearly, but it was positioned a bit east of its apparent location. Perhaps this was caused by the frontal zone being rather diffuse at 7 A.M. when the chart was drawn. By the time I reached the front at noon, it was better organized. There was the slightest suggestion of a trough on the 500 mb chart, and the temperature at 18,000 feet was about -7 C over the area. It had been $+16$ C at 9,000 feet so there was rather strong instability above 9,000 feet. That, plus the lifting in the frontal zone and a good supply of moisture, was enough to result in the line of showers.

SEPTEMBER—
HOUSTON TO BOWLING GREEN

The Perfect Warm Front

It was hot in Houston in late September. High temperature records were being set all over Texas and I knew from watching the TV on Wednesday night that there was a cold front to the west.

The synopsis hadn't changed when I checked with the FSS the next morning for a flight to the northeastern U. S. The map, as described on the phone, depicted a low pressure center in eastern Oklahoma with a cold front to the southwest and a warm front off to the southeast—the typical frontal model. The briefer suggested that the warm front was more active nearer the low, which is logical, and read the radar report to substantiate this observation. A broken area of thunderstorm activity was over Arkansas, with lines and clusters of storms,

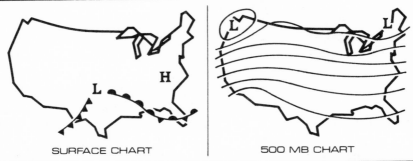

SURFACE CHART

500 MB CHART

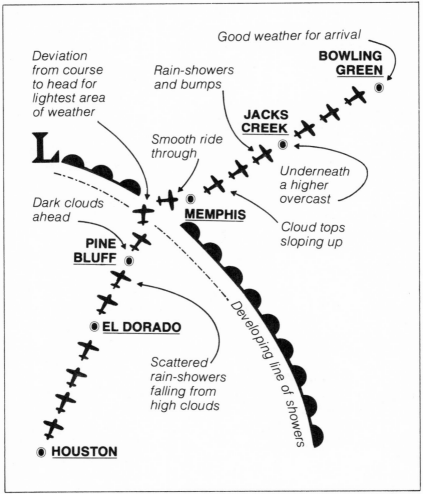

Good weather for arrival

BOWLING GREEN

Rain-showers and bumps

Deviation from course to head for lightest area of weather

JACKS CREEK

L

Smooth ride through

Underneath a higher overcast

Dark clouds ahead

MEMPHIS

Cloud tops sloping up

PINE BLUFF

EL DORADO

Developing line of showers

Scattered rain-showers falling from high clouds

HOUSTON

tops to 37,000 feet. The area extended east to Jackson, Tennessee, and southeast to the Mississippi border.

Little Rock was reporting a thunderstorm at 5 A.M.; all other reporting stations along the way had good VFR conditions.

Bowling Green, Kentucky, was our selected fuel stop, and the weather there was good and forecast to remain so.

The briefer made a strong recommendation that we go around to the south, to stay away from the depicted area of thunderstorm activity on the summary chart. I elected to go ahead and file for Bowling Green, though, on the theory that there would be plenty of time to examine the situation before reaching the warm front and to plan a deviation of minimum proportions. Too, it was early morning and, as often noted, reports and forecasts seem more fallible at the beginning of a day than at any other time.

Winds aloft were forecast to be rather light southwesterly, indicating that the low center wasn't a strong one. After leveling at 7,000 feet in the Cardinal RG, the groundspeed was calculated at 145 knots, about eight over the true airspeed. That verified the forecast wind of 240 degrees at 7 knots and was an indication that the pressure patterns were as the forecaster had expected.

The weather was beautiful as we proceeded northeastward, but when not far north of Houston we could begin to see a cloud shield to the north and northwest. The Stormscope was indicating activity quite a distance ahead. When control of our flight was transferred from Houston to Forth Worth Center, I asked about any depiction of weather on the traffic control weather. Fort Worth's area goes into southern Arkansas and sure enough they were showing some weather there. The controller indicated that nobody had been having trouble getting around the activity in the area, though, and we proceeded smoothly at 7,000 feet.

As we neared El Dorado, in south-central Arkansas, some rainshowers were visible ahead. The rain was falling from high clouds, based at about 10,000 feet. There was a low layer of broken clouds, with excellent visibility in between the two. It was a situation more typical of an occlusion, where the activity in a front is aloft and usually of not much problem to low level flights. I wondered if perhaps the whole low and frontal system might not be washing out, but there was still some indication of activity up ahead on the Stormscope.

After Fort Worth transferred us over to Memphis Center I started getting a better picture of the situation. The controller said there was activity to the southwest of Memphis that was forming into a line. The western end was about eighty miles to the northwest of Memphis and the line extended southeast-ward at least as far as Muscle Shoals, Alabama. Muscle Shoals is almost out of the Memphis area so the activity might well have gone on farther east than that.

The question of fuel supply vs. weather was reviewed at this point. My preflight estimate was for four hours and 35 minutes en route, six hours and 15 minutes fuel. That's a good reserve but Bowling Green has only a VOR approach and if their forecast was incorrect and the weather went to pot, I'd want to be headed for an airport with an ILS.

The general look of the situation suggested low ceilings. With the permission of the controller, I dropped off the frequency to check weather, but got no answer from Little Rock Radio on any of their published frequencies or from Memphis Flight Watch on 122.0. I returned to the Center frequency and the controller there furnished me with the necessary information.

Very interesting. Memphis was reporting 500 overcast and two miles visibility in rain. Jackson, Tennessee, about seventy-five miles northeast, had 600 scattered, 2,600 overcast, and three miles visibility in rain. Bowling Green was still all but clear. The winds aloft remained light, not indicative of rapid movement of the system, so it seemed reasonable to continue for a while on the original plan. It would be reevaluated again passing Jackson, Tennessee, which has a full ILS with minimums of 200 feet and a half mile visibility. The weather system wasn't likely to violate those minimums and Jackson became the alternate to Bowling Green in my mind. If Bowling Green weather started deteriorating, I'd just pull into Jackson with plenty of gas for contingencies.

As we moved along toward Memphis, the thoughts of fuel gave way to thoughts about the dark gray line in the sky that lay ahead. The Stormscope showed activity ahead and to the right. The approximate nature of the device's range information was circumvented by visual examination of the weather and by quizzing the controller.

The situation was defined as a broken line, about twenty-five miles ahead and twenty-five miles thick. We were heading 060. To the eye, there appeared to be a light spot on a heading of about 080. The Stormscope didn't agree. Nor did the controller who suggested flying a heading of 010 for forty miles and then going direct to Memphis. I settled for a heading of 030, which looked good both visually and to the Stormscope.

I was operating clear of cloud but it was apparent that I might get into a layer of clouds before actually reaching the line of weather. The best I could do there was aim for the low spots in the layer ahead to maintain visual contact for as long as possible.

Closer, it was possible to see a building cumulus to the left. The actual thunderstorm activity was to the right; the cell to the left had not yet matured but I had no desire to fly through it. The controller said that present position to the Memphis Vortac looked good and the Stormscope agreed. In eyeballing the area ahead, I picked the lightest spot, slowed to maneuvering speed, and flew a heading that would take the airplane toward that area. I flew into the cloud with everything set and running smoothly.

There was some rain and some very light turbulence in the line of showers. In a few minutes we were out on the other side. A Sigmet was issued for thunderstorms in the area soon afterward.

How to best pick a heading through such a line? In this case it was based on three things: visual, ground radar, and Stormscope information. Any two of the three would have been almost as good, and I suppose that it would have been reasonable to go with only one of these items of information if it strongly suggested a good ride.

Once the decision was made on a heading to fly, I felt committed to go on through. After we moved into the area of weather, the best way out would be straight ahead, even if the going got wet and bumpy.

It was interesting that this was in an area where isolated severe storms had been forecast on the "Today" program weather map. Perhaps that happened later in the day. We went by at noon, and the severe conditions just did not exist, again as shown by the three items of information.

Once on the other side of the weather, we were initially in the clear, on top of an overcast that appeared to slope upward ahead of us.

The slope soon rose to engulf the airplane in bumpy stratocumulus clouds as we moved along at 7,000 feet. I could see even higher tops ahead before we went on the gauges, and it was apparent there would be some wet and bumpy moments. These tops were not high enough to be thunderstorms but they promised a bit of tossing.

There was enough turbulence in the clouds to suggest slowing to maneuvering speed. There were no strong up- or down-drafts. It was just simple shear-type stuff. Even when flying through periods of heavier rain, there wasn't much up/down action.

In response to our query, the controller said that there were some rainshowers on up the airway, that nothing looked bad, and that airplanes had been moving through the area without complaint. In a bit, we flew out of cloud and were under a higher overcast, from which rain was falling. There were numerous clouds beneath, and in some spots the lower clouds merged with the base of the overcast above us. The rain was moderate at times, and there were a few moments of light turbulence.

Based on groundspeed calculations, the winds aloft went almost calm in the frontal zone, so it wasn't much of a front. We dropped back to 135 knots on one leg, a little of which might be attributed to our slowing down because of bumps.

We flew out of the weather area about seventy-five miles before reaching Bowling Green and found the weather there to be all but clear.

In retrospect, the light winds aloft probably told more about the relatively benign nature of the system than anything else. The low center was, I feel, just to the northwest of Memphis when we passed by there. The surface position of the warm front was west of Memphis. Flying through the area of thunder-showers and then into the upward sloping clouds was a good example of the textbook slope that is depicted when you look from the warm sector, through the front, and into the area ahead of a warm front. The temperature at 7,000 feet dropped 10 degrees C in the relatively short distance between the point where we flew into the sloping clouds and where we flew out the bottom of that cloud mass.

More strength in the system was suggested by the difference in temperature on the warm and the cold side, which was considerable. The National Weather Service did issue a Sigmet to cover the activity, but the 500 mb chart showed a definite lack of upper level support for really serious stuff. But, as is true in all weather situations, the measure is in the total of all factors, not in one or two isolated items.

Flights: Fourth Quarter
October, November, December

This quarter starts with the beautiful color of fall in the north and, before it ends, the beauty of the season has moved south. And then the leaves all blow away, making the transition to winter complete.

As far as flying weather goes, fall seems to be characterized by a lot of VFR weather and then by a lot of IFR weather. Dry and then wet—it has always seemed to go from one extreme to another, with strong variations from month to month. For example, one year I flew forty-five hours in October with but one hour and 40 minutes of that in actual IFR conditions; the following month, November, I flew sixty-seven hours with a full twenty-four hours of that in actual IFR conditions.

Over the years, the time spent in actual IFR conditions has been pretty much the same for the first three-quarters of the year, and has increased dramatically in this last quarter, usually in November. In fact, averaged over eight years, actual instrument time has run half again as high for this quarter as for any of the other three.

Fall turns to winter more gracefully than winter turns to spring, though, and severe storm systems aren't the big bother during this time of year. Usually when the weather is bad, it's just rain, with the big thunder earlier and the frozen stuff coming later. But anything can happen.

It's always interesting to see when winter really asserts itself. Most of the time, the real misadventures with snow and ice wait until January, but in some years and some parts of the country, December offers a good dose of wintertime.

In a nutshell, this quarter is usually downhill. October is nice, November can show the beginnings of winter, and by December the days are shortest and the deep chill is upon us.

OCTOBER—TRENTON TO FLORENCE

An Early Fall Coastal Low

The storms that form on stationary fronts and then move up the east coast of the U. S. are always interesting. This particular situation wasn't expected to amount to much, but it developed into an unusually enthusiastic storm for the time of year. It baffled the forecasters until they recognized the full development of the system.

In describing the synopsis for the flight from New Jersey to Florida, the briefer positioned the low off Cape Hatteras with a stationary front extending to the north and a cold front trailing to the south. A four-hour-old radar summary chart showed a line of thunderstorms over the west coast of Florida moving eastward at 15 knots and a scattered to broken area of rain over the Carolinas. The Philadelphia weather was clear, Richmond and Raleigh-Durham were okay, and Florence, S. C., was reporting 300 scattered, 2,000 broken, with five miles visibility in light rain. I was going to Kissimmee, Florida, and on the basis of a reasonably good forecast (1,400 broken, 8,000 overcast, five miles visibility), I chose Florence as a fuel stop. The briefer noted that anything to the west would be good if things went sour along the coast. In fact, he said that except for weather connected with that low, the whole country was clear. And we were even slated for a tailwind. I didn't ask for any information off the 500 mb chart and thus missed what was to become an important feature of the day's activity.

Conditions were spectacular as we left Trenton. It was one of those perfectly clear mornings. The New York skyline fifty

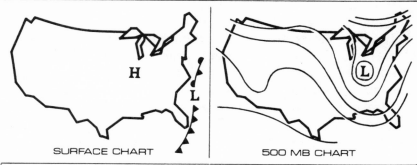

SURFACE CHART | 500 MB CHART

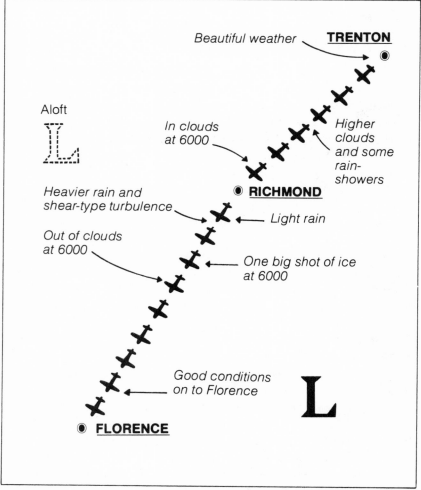

Beautiful weather — **TRENTON**

Aloft

L

In clouds at 6000

Higher clouds and some rain-showers

Heavier rain and shear-type turbulence

● **RICHMOND**

Light rain

Out of clouds at 6000

One big shot of ice at 6000

Good conditions on to Florence

L

● **FLORENCE**

miles northeast, and the Philadelphia skyline forty miles south-west, were sharply visible. The cloud shield to the southwest was plain to see, though, and we hadn't flown far before some showers in southern Jersey were visible.

I had filed for 6,000 feet to take advantage of a forecast northeast wind at that level. The breeze aloft was a little more easterly than northeasterly but the groundspeed was still in excess of the true airspeed and that's never anything to complain about.

There were a lot of complaints coming from pilots at higher altitudes. There was apparently significant turbulence from 9,000 on up. One airline pilot said that he had been jostled all the way up to 35,000 feet (Flight Level 350). That would certainly be expected in this sort of weather situation. The low level circulation around the surface low off the coast was reasonably strong—15 to 30 knots—and the winds shifted from the east (the effect of the surface low) to the south at 9,000 feet and up. There was plenty of wind shear involved, because of the change in direction with altitude, and shear means turbulence. Winds at 18,000 feet were southwesterly at 65 knots so the velocity did increase rather rapidly with altitude as the wind changed direction.

At 6,000 feet I was beneath the base of a higher overcast but the view ahead suggested that this wouldn't last for long. I could actually perceive the slope of the clouds. There was no indication of thunderstorm activity along the route and I was looking forward to a nice smooth ride, even after punching into the clouds.

Along about Richmond, I moved into cloud at 6,000. There was some light rain but the flying was smooth south of Richmond. Then things took an interesting turn. The rainfall rate increased dramatically, and there was some turbulence. No updrafts and downdrafts, as in convective activity, but more the sharp jabs and moderate airspeed fluctuations of wind shear. I thought about a different altitude, but higher would have been with less favorable wind and lower didn't seem to offer much better conditions.

The temperature had been several degrees above freezing but it dropped rather suddenly in connection with a period of light to moderate turbulence and very heavy precipitation.

Some ice formed—it looked as if someone had thrown a bucket of water at the airplane and it had frozen—but the temperature returned to above freezing in a moment or two and the ice left as quickly as it had come. The temperature often does drop as precipitation begins and brings cold air down with it, but this was the clearest case of that phenomenon that I had ever seen.

All the time I was flying in heavy precipitation it also seemed that I was close to the cloud tops. There was light instead of the darkness that you find in heavy precipitation in thunderstorm areas. And when we flew out of the backside of the weather and into an area with lower scattered clouds and a high overcast above, the higher clouds appeared to be quite thin. There were still some rainshowers around, but the weather at Florence was quite good. It was cold, too—+9 C degrees at high noon on October 13 in Florence, S. C., is a touch chilly.

The conditions encountered were quite reminiscent of what you might find flying through a warm front, but that wasn't the situation that was presented before our flight. The weather map showed no fronts along the way and the low was to the east. You don't usually get such a wet and bumpy ride well to the west of a low pressure area but in this situation the explanation is found on the 500 mb chart. Note the closed low aloft, inland, to the west of the surface low. There was certainly activity between the two as air moved around the surface low and up into the low aloft, and that was what caused the weather to the west of the surface low.

OCTOBER—TERRE HAUTE TO TRENTON

Weak Cold Front

If there was any question about weather for this flight, it was in relation to icing. Terre Haute was a fuel stop and I had accumulated some ice descending to land there. In fact, a stratocumulus deck that extended from 4,000 up to 9,000 had given clear warning that it would not tolerate extended flight in its midst.

The stratocumulus deck was behind a rather weak cold front between Terre Haute and Trenton. All stations were reporting

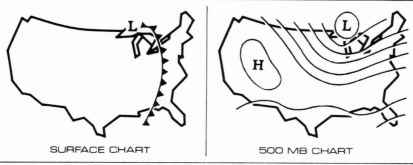

SURFACE CHART   500 MB CHART

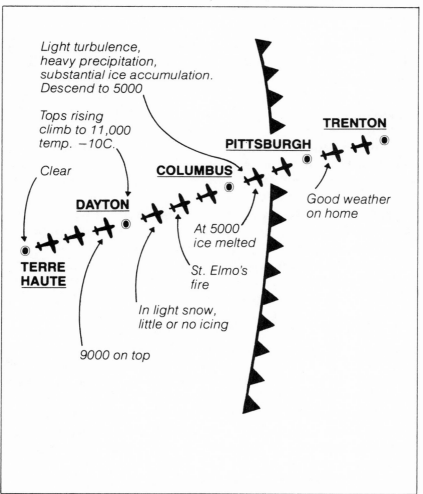

Light turbulence,
heavy precipitation,
substantial ice accumulation.
Descend to 5000

Tops rising
climb to 11,000
temp. −10C.

Clear

TRENTON

PITTSBURGH

COLUMBUS

DAYTON

Good weather
on home

At 5000
ice melted

**TERRE
HAUTE**

St. Elmo's
fire

In light snow,
little or no icing

9000 on top

4,000 foot ceilings or better and good visibilities. The radar summary chart positioned a few echoes in Ohio but the tops were shown as only 20,000 feet. Winds aloft were favorable and the destination, Trenton, was far enough ahead of the front to have a forecast of excellent weather into the night. Surface temperatures all along the way ranged from +10 C up to +15 C well to the east of the cold front.

The sun would set about an hour into the four-hour flight, so I opted for an IFR flight plan. Based on the briefing, VFR would be possible but, one, I don't like to fly VFR at night and, two, anytime there is a front on the map it is best to suspect the en route weather, regardless of good reports and forecasts.

The weather had cleared when I left Terre Haute. The stratocumulus deck had moved eastward, and I was soon skimming its top at 9,000 feet. There were low spots and high spots and by deviating a bit to either side we were able to stay clear of cloud, and thus clear of icing.

As it got darker, it was difficult to stay out of the clouds, though, so I got a clearance up to 11,000 feet. That was on top initially. The temperature was −10 C, certainly not too cold for ice in the tops of stratocumulus clouds. The center controller responded to an ice query with a report that the only aircraft reporting ice were south of Indianapolis and that it was reported as extending up to 17,000 feet in that area. A pilot fifty miles ahead heard our request and volunteered that he was encountering very light rime ice at 11,000.

The Center did volunteer that there was some precipitation return ahead, on the airway, but that airplanes had been flying through it without comment.

It being dark, there was no way to tell whether the tops of the clouds beneath had risen above 11,000 feet before we flew under higher clouds, but it was certain that the situation had changed. It was snowing lightly, but we weren't spending a lot of time in thick cloud. The snow was illuminated by the strobes, but the blinding flashes you see when strobes are on in cloud were few and far between. There was a little ice on the wings from earlier in the flight, when it took a few minutes to get approval to climb to 11,000, but only the slightest trace of new ice appeared. In all, it was rather comfortable.

There was plenty of electricity, though. St. Elmo's fire illuminated the propeller tips, and every time I got my hand within a few inches of the windshield, it sparked at me. St. Elmo's is allegedly harmless, but it has always seemed to me some sort of admonition from nature. *The good Saint takes this opportunity to remind you that you are a mere mortal, trespassing in his electrical field.* Zap. And on it went, droning in the night at 11,000 feet. Occasionally the moon would be dimly visible above and lights on the ground would be dimly visible below.

Taking stock about seventy-five miles west of Pittsburgh, I came to the conclusion that I had passed through the front and was all but home free. The Pittsburgh weather was good—scattered clouds at several levels and excellent visibility—and I was of the opinion that I was well past the surface position of the front. That bit of intuitive reckoning, however, turned out to be incorrect.

I flew into an area of light turbulence and heavy precipitation about fifty miles west of Pittsburgh. The ice accumulation in this area was quite rapid. Rapid accumulation with the temperature at −10 C is restricted to cumulus clouds. As a result, I could only conclude that I hadn't really passed through the front back there, I was passing through the frontal zone now.

The tailwind was nice at 11,000 feet, the oxygen bottle still had some in it, and I would have liked to stay high. But the principal rule in dealing with ice is to do something about it, so I told the controller that I was accumulating ice rather rapidly and would like a lower altitude. He obliged and I had to descend to 5,000 feet to shed the quarter to half inch of ice that had accumulated. Of course, by the time the descent was completed I had flown out into the clear, on the east side of the front, and I couldn't help but think that it would have been better to stay at 11,000, ride it out, and sacrifice 5 knots to the accumulated ice to keep 15 knots better tailwind. But that was hindsight.

The flight was a good reaffirmation of the significance of fronts, too. Everything indicated that there was not much weather in the front, and that I would have been able to go VFR if desired. Yet that area of light snow aloft was 150 miles through, and there was heavy precipitation for about twenty miles. Even with good ceilings, the visibility beneath couldn't have been very good through the frontal zone.

Fronts are not impenetrable walls in the sky, as I thought they were when I first started flying. But they are warnings that a good chance exists of weather being encountered in that area. In this case, the 500 mb chart supported the frontal activity with a rather strong, closed low just west of the area where I encountered the front.

NOVEMBER—SANDUSKY TO ROCKFORD AND MINNEAPOLIS TO CLEVELAND

Double Dipping the Front

The early November trip from the east coast to Sandusky, Ohio, a gas stop, was uneventful. I thought, though, that there would be some weather to the west of Sandusky. The morning weather map (on the "Today" program) had shown a low in eastern Oklahoma, moving northeastward along an all but stationary cold front across central Illinois. Also, as I flew westward there had been a rather dramatic change in pressure. The altimeter setting at Trenton, N. J., was 30.37; it was 29.91 at Chicago when I checked the weather before departing Sandusky for Rockford, Illinois.

The reported weather along the way was not too bad. South Bend had 2,600 broken, 4,400 overcast, and two and a half miles visibility in light rain. My destination, Rockford, was reporting 700 overcast, three miles in fog. The forecast for Rockford read 1,200 broken, 3,000 overcast, five miles in fog with a chance of lower conditions in thunderstorms. Surface winds over the area were fresh and southerly. Winds aloft were strong and southerly. The radar summary chart showed a broken area of thunderstorms and rainshowers, tops to 34,000 feet, along the Indiana and Ohio border. This was apparently in a rather diffuse frontal zone.

As I was flying at 8,000, a higher cloud deck appeared overhead not far west of Sandusky, followed shortly by showery cumulus types. The bases of these were down around 3,000 to 4,000 feet. The controller said that there was some weather ahead at one point and offered a vector to the south. At this time the advertised south wind at the cruising altitude had shifted around to a strong southwest-

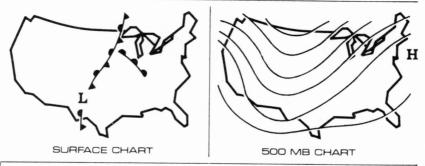

SURFACE CHART | 500 MB CHART

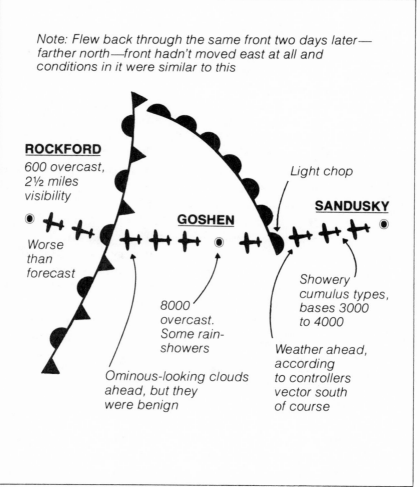

Note: Flew back through the same front two days later—
farther north—front hadn't moved east at all and
conditions in it were similar to this

ROCKFORD
600 overcast,
2½ miles
visibility

Light chop

SANDUSKY

GOSHEN

Worse
than
forecast

8000
overcast.
Some rain-
showers

Showery
cumulus types,
bases 3000
to 4000

Ominous-looking clouds
ahead, but they
were benign

Weather ahead,
according
to controllers
vector south
of course

erly flow of about 35 to 40 knots. There was some light chop in heavier showers.

Near Goshen, Indiana, the situation changed to about 8,000 overcast with scattered rainshowers. I made a note that it seemed as if a little front had been located in this area. I changed the cruising altitude to 6,000 and later to 4,000 in an attempt to minimize the effect of the wind as well as to avoid some turbulence close to the bases of the overcast.

The nearest thunderstorm activity was well to the southwest from all the reports I got, but as it got dark some clouds ahead took on an ominous appearance that suggested the possibility of problems. Nothing came of it, though.

The Rockford weather this evening was lower than forecast, at 600 feet and two and a half miles visibility. The forecast was probably based on the front moving a bit, whereas it actually remained stationary or perhaps even backed up a little.

This same front was in roughly the same location two days later for an eastbound flight from Minneapolis to Cleveland. Again it was spawning some rain, but there were no thunderstorms and the reported weather wasn't bad when I checked rather early in the morning. The worst report along the route was at Eau Claire, Wisconsin, which had 1,000 broken, 6,000 overcast, and six miles visibility in rain.

My cruising altitude was 7,000, and as I flew into the frontal zone there was another reminder that fronts can always cause weather even when they have been stationary for a long while, and even though initial reports in the morning can be good. Several airplanes had missed approaches at smaller airports in the area, and a rather general area of moderate rain had developed all along the frontal zone and had become rather broad. The effect of convergence was bound to be present, too, because the Minneapolis winds aloft showed 280 degrees at 20 knots at 9,000 feet while Cleveland was showing 170 at 15 at the same altitude.

I began to wonder a little about how weak the front might be as my little airplane moved through the front. For a few moments the turbulence became rather enthusiastic and the rain heavy. Then it abated and I was on the east side of the front.

This bit of weather was the type that is ideal for IFR flying. The freezing level was high even though it was November in the northern part of the country. The frontal zone contained a lot of rain but not much turbulence. And the weather didn't often go below IFR minimums over a wide area. A look at the upper level, 500 millibar chart, shows some reason for the relatively tame nature of the system. The flow over the area was not extremely strong aloft and it was on nearly a straight westerly flow.

For VFR, though, it could have been a particularly treacherous situation. Reports suggested that VFR would have been possible, but there were enough areas of heavy rain with lower layers forming in the heavy rain to preclude VFR operations. There would likely have been a rather solid line of impossible conditions for any VFR pilot trying to go east or west through the frontal zone.

NOVEMBER—CLEVELAND TO TRENTON

A Bad High

We don't often think of a high pressure area as creating an area of bad weather, but it happened one night, on the continuation of the flight just related.

The briefing at Cleveland wasn't of much value because the FSS computer was out of service and I could get only snatches of old information. The total of useful information was the synopsis, a high off the Massachusetts coast, plus the Phillipsburg, Pennsylvania, weather at 300 obscured and three-fourths of a mile visibility, Philadelphia weather at 500 broken, 900 overcast, and a Teterboro forecast of 3,500 scattered and 10,000 broken. The Teterboro forecast and the Philadelphia weather didn't really go together, and I wondered if perhaps they had picked up some reports from a previous day to use in the absence of information from the computer. Winds were southwesterly at Cleveland and southeasterly along the coast. The briefer didn't mention them, but from watching the "Today" program that morning I knew that there were supposed to be a couple of lows to the south. Low to the south and high to the north means east wind and east wind along the Atlantic coast often means trouble.

This was to be a night flight, and I was more than a little concerned about an alternate for Trenton. The FSS person

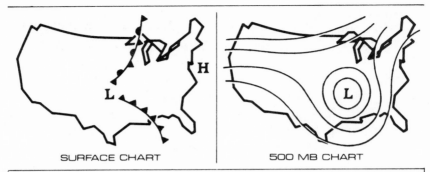

SURFACE CHART

500 MB CHART

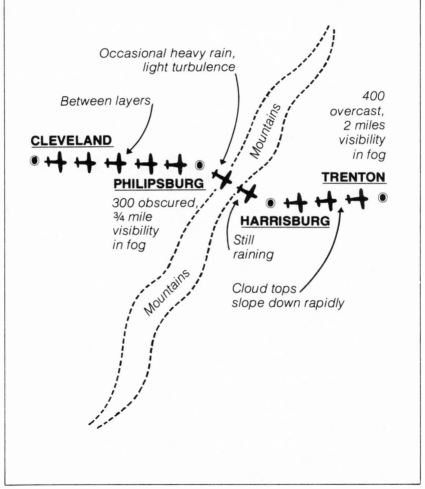

Occasional heavy rain,
light turbulence

Between layers

400
overcast,
2 miles
visibility
in fog

CLEVELAND

PHILIPSBURG

TRENTON

Mountains

300 obscured,
¾ mile
visibility
in fog

HARRISBURG

Still
raining

Cloud tops
slope down rapidly

Mountains

gave that good Teterboro forecast and I promptly used it as an alternate on the flight plan but in my heart I knew that there would have to be other alternates. First, I would check weather aggressively and not go too far from Cleveland (where the weather was good) unless I was able to get a better picture of the coastal weather from an FSS along the way. Second, after I got weather along the way I would recompute alternates based on the ILS airports in the area, such as Allentown, Reading, and Harrisburg, Pennsylvania. Third, I would keep careful track of winds and develop a point of no return to the good weather in the Cleveland area. I rather thought that point would be Trenton—meaning I could go there, miss the approach, and then return to Cleveland with adequate reserves —but there's nothing like being sure.

The weather was layered to the east of Cleveland and the flight was quite beautiful. The lights of towns below illuminated the lower layer, and airplanes could be seen zipping around between layers.

I was fully expecting a nice smooth flight but the very basics of the elements were opposed. It was logical, too. The easterly flow along the coast was moving against a basic southwesterly flow, the moisture supply was plentiful, and the mountains gave some lifting. Even though no reporting station gave anything other than light drizzle, I had some brief moments of heavy rain and light turbulence. It was nothing to write home about, but based on the sequences and forecasts that I gathered as I flew along, there shouldn't have been any rain at all.

The rain persisted until I was near Trenton. There the tops sloped down very rapidly and I was on top at 9,000 feet for the last portion of the flight. The Trenton weather on arrival was 400 overcast and two miles. Once I had gotten my first bit of weather after leaving Cleveland, there was not much doubt about continuing because all airports with ILS approaches were comfortably above minimums.

In reconstructing the weather from this flight, it is interesting to consider how nice it would have been to have a cross-section of weather before takeoff, to know at least what to expect. Had I applied the basics to the scanty information I had before takeoff, I could have had some idea of this; given a couple of more facts, I could have had a really clear picture had I made the effort.

The easterly flow meant clouds over the coast and it was certainly logical that the tops would become higher over the mountains and that there could be some rain. After all, an easterly flow would be full of moisture and it was bumping against a southwesterly flow that would also be carrying moisture from the Gulf of Mexico. The fact that the flows were weak —20 knots was the strongest wind around at 9,000 feet, and the 500 millibar chart reflected very little flow and not particularly cold temperatures aloft—shows why wet and just a *little* bumpy was the order of the day.

NOVEMBER—LITTLE ROCK TO WICHITA

Cold Fronts Take All Forms

Fall turns to winter slowly and in a series of cold fronts. This front, in November, had the coldest air of the season behind it. The low was out in Colorado and the front made a big circle to the east and then to the south and southwest. When the circulation around a low is so strong that the cold front moves around the low instead of just trailing away to the southwest, it is a sign that the low is deepening, strengthening, and that wild things are in store ahead of the low when it starts moving. My problem this day was in dealing with the front a substantial distance away from that deepening low, though, and it didn't appear too bad.

Fort Smith, 110 nautical miles up the line, reported 3,000 scattered, 6,000 overcast, and was forecasting 2,000 overcast with a chance of heavy thunderstorms. Tulsa was down to 400 overcast and a mile and a half in light rain. Wichita was reporting 700 overcast and six with a forecast of 500 and one in light rain with a chance of thunderstorms. The Wichita surface wind was out of the southwest and light; it was blowing like a bandit from the southeast in Little Rock.

The winds aloft forecast called for a very strong southerly flow, becoming lighter by Wichita. Given the situation, I planned to watch actual winds closely. They would give a good indication of my position in relation to that front.

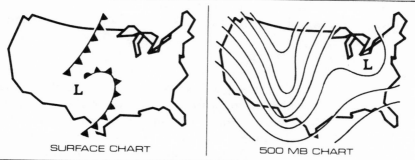

SURFACE CHART   500 MB CHART

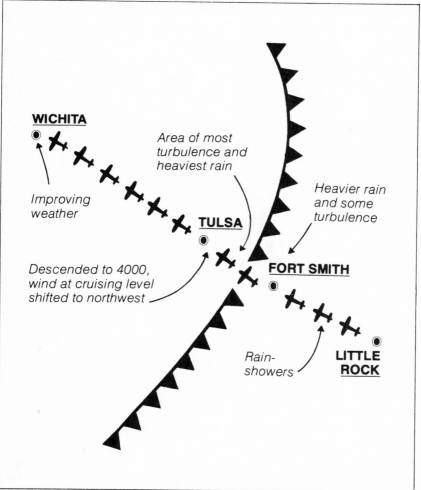

WICHITA

*Area of most
turbulence and
heaviest rain*

*Improving
weather*

TULSA

*Heavier rain
and some
turbulence*

*Descended to 4000,
wind at cruising level
shifted to northwest*

FORT SMITH

*Rain-
showers*

**LITTLE
ROCK**

The radar summary chart depicted a broken area of rain-showers and thunderstorms over eastern Oklahoma and southern Kansas.

I started out at a cruising level of 8,000 feet. Perhaps that was a little high for the situation, but given the Boston mountains to the south, and the strong southerly flow, it seemed a good idea to use some altitude to soften the mechanical turbulence in the area. I was very familiar with the route and had been bounced around in there enough to know how it worked.

It took a 20 degree drift correction to track the airway, bearing 297 degrees, and the controller said that our groundspeed was 150 knots, reflective of a 10 knot tailwind component. I didn't run that through a computer at the time, but I did think that the wind must be a little east of south to get those results. With a computer, I found the southerly direction to be correct, but the velocity was about 15 knots in excess of the forecast 35. This shouldn't have been any surprise because winds generally increase as you get closer to a frontal zone. That is, after all, the area of climax. When they forecast wind, it is done for the general area. Any general increase in the frontal zone seems never to be covered.

The air traffic controller said there was a large area of weather ahead and offered a vector around one heavier spot that he was painting. The Stormscope indicated no thunderstorm activity ahead at all. There was some off far to the south but none to the northwest.

In the everlasting search for smoother air at a lower altitude, I requested 6,000 feet before reaching Fort Smith. This was approved and, while it was still bumpy there, it seemed smoother.

One of the ways to study a weather system when you fly through it is to note the type of turbulence encountered. This day we had experienced some up and downdrafts when on the lee side of the mountains. That was to be expected. Once away from the area of mountain effect, the bumps were sharp jabs with little vertical effect. The airspeed would vary perhaps 10 knots but that was all.

More interesting things were ahead, though. The controller said that the weather return on the scope was pretty solid, then he offered vectors around the heaviest areas. Next he said that the airway was as good as any way. The rain was very heavy at times and quite a bit of vertical activity was encountered in the heavier rain. I was trying to measure this and found that if I maintained airspeed I'd get as much as 1,200 fpm rate of climb followed eventually by slightly lower rates of sink. If I responded to the vertical action by maintaining attitude and pulling power back to hold altitude as closely as possible, I'd get rapid airspeed increases of as much as 20 knots. Still, except for a few extra enthusiastic pokes, the turbulence couldn't have been classed as more than moderate.

Before reaching Tulsa, I got further descent to 4,000 feet. At that altitude we soon flew out of cloud and into an area of rain falling from an overcast based at probably 5,000 feet. I could almost sense the frontal slope, from just over my head at this point back to the surface at a point some miles back. And about this time we got the most pronounced turbulence of the trip. The groundspeed went to zilch, too, dropping from 150 knots to 110 knots. There was no drift with the new wind, so it was northwesterly. That meant that we had flown from an area with a 50 knot southerly wind into an area with a 30 knot northwesterly wind, with the transition or frontal zone being about ninety nautical miles wide. No wonder it was a little bumpy. But remember, the turbulence was not caused by thunderstorms. It was a product of converging air and wind shear. The difference is important to understand, lest you approach the situation with the wrong attitude.

Why were there no storms as such in this frontal zone? For one thing, there was no strong and basic upper level support to encourage storm development. The surface low pressure center was getting organized and gaining strength rapidly but it was far away. The action aloft was southerly, and no identifiable low was around to give support. The trough aloft extended far down into Mexico and the temperatures aloft were relatively warm over the area. It was $+20$ C at the surface and -12 C at 18,000 feet, for a lapse rate of 1.8 degrees C per 1,000 feet, or a little less than required for real instability.

The tendency of a surface low to advance northeastward (or northward in this case) along the east flank of a trough aloft meant that the Little Rock to Wichita leg was in an area well clear of the path of the low. If the tip of the trough at the 500 mb level had been just west of, or right over, that path, things might have worked differently.

In such a general and wide frontal zone, with widespread rain, the controller was probably correct in saying that the airway was as good as any way. The variation in rainfall rates wasn't enough to distinguish on a traffic control radar scope. It would probably have been difficult with the finest in weather radar equipment. The Stormscope told me that there was no lightning in there, thus no thunderstorms, which was useful information—but it did nothing to calm the turbulence that I did encounter.

The general characteristics of this front were not suggestive of severe weather. A broad cold frontal zone suggests that it is rather diffuse at that time and place. This one was moving slowly because it had all but outrun its strong low pressure center and had to pause for some regrouping.

It was still raining when we landed in Wichita and the surface wind was out of the northwest at only 10 knots. But in tracking the ILS to Runway 1, it was obvious that there was a very strong northwesterly flow aloft. The wind later picked up to 40 knots. The Wichita weather cleared beautifully the next day, but after gathering strength and regrouping, the low in Colorado took off and brought a real November blizzard to Nebraska, Iowa, and Minnesota.

NOVEMBER—TRENTON TO ASHEVILLE

North of a Stationary Front

The weather along the east coast had been bad for days in late November. There were rumblings of a big cold air mass to the west, but it had remained soggy and relatively mild in our area. Then a cold front moved barely through, became stationary, and a succession of waves developed on the front and moved to the northeast. There was snow in northern New Jersey the morning of the flight. The continuous weather broadcast mentioned an icing layer between 1,500 feet and 3,000 feet, plus icing above 9,000 feet.

The FSS (and the "Today" program) positioned a low center in southern Indiana. A warm front extended to the east and was

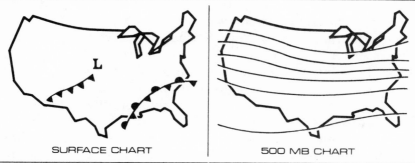

SURFACE CHART   500 MB CHART

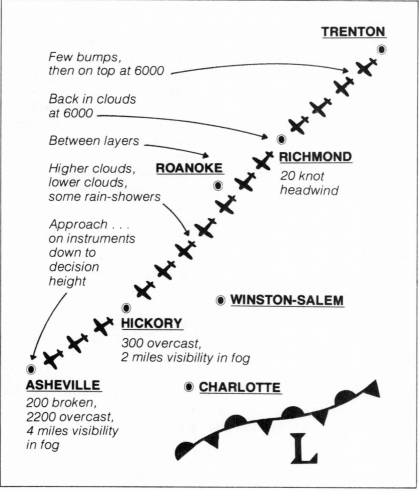

TRENTON

Few bumps,
then on top at 6000

Back in clouds
at 6000

Between layers

Higher clouds,   ROANOKE
lower clouds,
some rain-showers

RICHMOND

20 knot
headwind

Approach . . .
on instruments
down to
decision
height

WINSTON-SALEM

HICKORY

300 overcast,
2 miles visibility in fog

ASHEVILLE   CHARLOTTE

200 broken,
2200 overcast,
4 miles visibility
in fog

L

just south of Trenton, N. J., the point of departure. A cold front trailed to the south. There was some thunderstorm activity in South Carolina and Georgia. Surface winds were light everywhere, from the northeast on the north side of the front and from the south on the south side of the front. Everything was forecast to remain relatively stationary. Winds aloft were southwesterly and in the twenties; temperatures aloft were forecast to be quite warm. The best weather I wrote down for the entire route was Philadelphia's 600 overcast with four miles visibility. Most others were 200 and 300 feet. The destination, Asheville, N. C., had a forecast of 300 overcast, three miles visibility, briefly a half mile visibility.

The search for an alternate took a few minutes, but I finally found Knoxville's forecast of 1,000 overcast with three miles visibility for late in the afternoon, when I'd be getting there. I took that as an alternate even though there was little desire to go there. It's across the high mountains from Asheville.

There were three of us in the airplane and, as usually happens, the windows fogged over on the inside as we taxied out. The surface temperature at Trenton was +5 C and warm bodies in cold airplanes plus high humidity will do that every time. It adds some atmosphere to the beginning of an IFR flight.

The ceiling was as low as advertised and I was in thick clouds at the initial clearance altitude, 3,000 feet. Somehow it looked brighter above than I expected, though. From the briefing, I thought we'd probably have layered clouds up to well above 10,000 feet all the way to North Carolina. At 4,000 it was even brighter, and as I flew through some bumps at that level I thought to myself that we must be moving through the rather diffuse warm frontal zone. At 5,000 I was all but on top, and at 6,000 I was on top. It was clear above but there were clouds to the southwest.

Our southwestward progress suggested that the winds were not as strong as forecast. I was being rather gentle with the power, to maximize endurance, but the groundspeed was still about 120 knots. That suggested about 15 on the nose.

As I checked the weather, the thought of maximizing endurance became more desirable, too. Virtually everything remained at 200 to 400 feet, visibilities about one mile.

When flying in changing weather situations, the challenge is to pick the best path. This day, there was no question about path. No storms stood in the way. Nor was there any challenge of figuring out what the weather was doing. It was dead, draped over the route, unchanging. This tends to be frustrating in the sense that there is never good news or bad news. If nothing changes, there is no suggestion that it is getting better, and that it is wise to continue. And there is no clear-cut sign that an alternate plan should be put into action.

Instead of considering Knoxville as my real alternate, I was thinking more in terms of all the airports along the route of flight with instrument landing systems. I just didn't relish going to Asheville, missing an approach, and then flying across the mountains to Knoxville. I would much rather try something in the Carolinas. Any solid hint that I couldn't make Asheville would send me to Winston-Salem, or Greensboro, or Charlotte, or Hickory, or Greer.

A couple of hours before I got to the Asheville area, I got a good report. Asheville had gone to 300 scattered, 1,800 broken, 5,000 broken, and six miles. Was something moving? Perhaps, perhaps not. All my ILS airports along the way were still barely clinging to minimums. The best was 500 feet, the worst 200 feet. The one that I really wanted to consider my alternate, Greer, S. C., was the worst at 200 feet and a half mile.

The next hour's weather painted the darkest picture of all. Asheville was down to 200 broken, 2,200 overcast, and four. The decision height is 250 feet, so that would be very close. Greer, the alternate that I really wanted, was down to 200 overcast and a quarter mile. Knoxville was still good, though.

I was close to Hickory, N. C., when this weather was obtained. Hickory, about seventy miles from Asheville, became an alternate in my mind. The reported weather was 300 overcast and two. The decision height is 250 feet, the minimum visibility one mile. Even better was the fact that I had heard some airplanes shooting approaches there and they were making it.

A recalculation of fuel and reserves showed that I could go on to Asheville, make an approach there, and go on to Knoxville and land with an hour's fuel in the tanks. Knoxville was quite good. I could also land at Hickory, at hand, and get some more fuel. Doing that would mean reaching Asheville after dark, though, and weather usually deteriorates after dark. No, Hickory really didn't offer very much. The logical thing was to go on to Asheville, with Knoxville as an alternate. When I asked the controller if they were making approaches into Asheville, he said he hadn't heard of anyone missing yet. That was a good sign.

When flying in such a weather situation, there is always some comfort in getting in contact with the approach control facility at the destination airport. The person there *knows* what the weather is, and knows whether pilots have been having trouble with the approach. When I got to the Asheville controller the good news was that approaches were being conducted successfully. The bad news was that the ceiling was now 200, variable 300. If that assessment of the ceiling were accurate it meant that there might be a runway in sight at the 250 foot decision height, or there might not. And, if there was not, the temptation would be to try another approach. And another. And how much gas did you say was in the tanks? I resolved to try two approaches maximum and then go to the alternate.

There was a Piedmont YS-11, an Aztec, and a Baron ahead of me. This resulted in some vectoring that probably added ten minutes to the flight time. That meant that I should delete that second approach in case of a miss. Drat. But you do have to take both the breaks and your turn in traffic.

Because the approach was critical, I planned ahead as much as possible. How would the winds change as I descended? The wind at 6,000 feet was southwesterly. Asheville was reporting a south wind at 10 knots, which would be downwind on the ILS to Runway 34. Thus there shouldn't be much shear effect. But the higher groundspeed caused by the downwind component would require a higher than usual rate of descent to track the glideslope.

The airplanes ahead were apparently going in okay, and my turn came. It was smooth on the ILS down to the tops of the lowest layer of cloud. Glancing briefly at that broken layer from above, it was difficult to imagine that there would be ceiling beneath it, but I continued.

The needles got off a bit when the light turbulence came. Apparently there was some shear. I was close to the decision height, a small correction was made, and when there were fifty feet to go to DH I pushed the prop control in and cocked my throttle hand. Just as the "missed approach" thought came to mind, the runway and its lights appeared ahead. Truly down to the wire.

An interesting thing happened later that evening. Instead of going to pot as weather often does after dark, it all but cleared. Still later in the evening it went back down, worse than ever, to a very dense fog.

In studying weather charts later, I found no identifiable warm front south of Trenton. Instead, there was a stationary front just south of Asheville. This was rather obvious during the flight. I certainly wasn't in a warm sector at any time. Aloft, the only trough was out to the west, brewing up a low and a cold front for a few days later.

When stationary frontal systems create widespread areas of bad weather, there just isn't a lot going for you other than a good alternate and a lot of fuel.

NOVEMBER—ASHEVILLE TO TRENTON

Low at the Destination

This was a reverse run of the preceding flight, flown a few days later. While we were in Asheville, the weather had remained cloudy and mild, but on Friday night before our Saturday morning departure there had been an apparent frontal passage, complete with thunderstorms.

The weather was checked before daybreak, and offered a few interesting possibilities. Asheville was almost clear but with occasional snow showers blowing by. Winds were 18 knots gusting to 27. Winds aloft at ridge level were high—northwesterly at 43 knots at 6,000 feet—so there would no doubt be turbulence. There were two lows on the map, one north of Buffalo and one southeast of the New York City area. En route weather was forecast to be good to about Richmond. The forecast for the Trenton area on arrival called for 1,000 overcast and three miles visibility with light rain, light snow, and fog. Surface winds were forecast to be northwesterly at 15 with gusts to 30. The surface temperature at Asheville was −5 C; it was +7 C at Trenton. Winds aloft were forecast northwest-

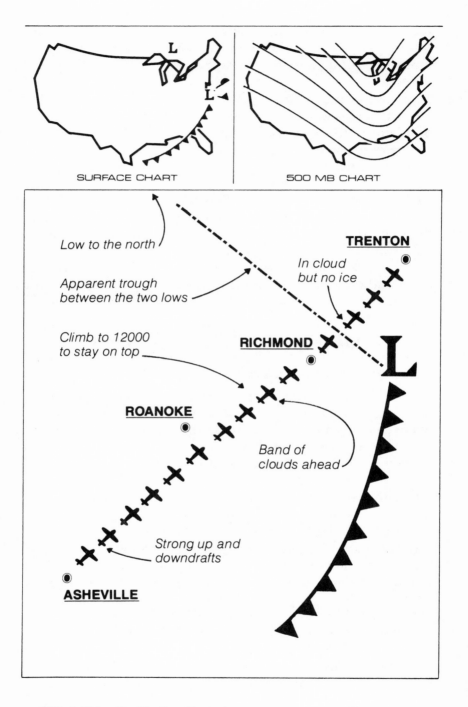

SURFACE CHART

500 MB CHART

Low to the north

Apparent trough
between the two lows

Climb to 12000
to stay on top

TRENTON

In cloud
but no ice

RICHMOND

ROANOKE

Band of
clouds ahead

Strong up and
downdrafts

ASHEVILLE

L

erly and strong to begin, shifting to southwesterly and strong later on in the flight.

There were the usual mentions of icing, and you do have to consider this in such conditions. It struck me that the best plan would be to go high, to minimize turbulence. Too, the temperatures aloft were forecast to be very cold (-18 C), and that should preclude icing in the cloudy area starting near Richmond. High enough would probably be on top anyway. When near Richmond it should be possible to determine through pilot reports how much icing would be gathered on a descent through clouds. If the answer was bad, then there would be the relatively clear area behind, where a descent could be made clear of clouds, and a stop could be made for regrouping.

In mountains, wind is weather and I must admit that my initial thoughts were more on how to minimize turbulence getting up and out of the Asheville area. A Sigmet called for occasional severe turbulence below 12,000 feet over rough terrain, and the challenge was avoiding the "occasional."

The air in which we fly takes on strange properties at times, and it did this morning as we left Asheville, nestled as it is down among the mountains. There were no really big bumps, but the air was busy. It was a reminder that you wouldn't want to fly directly downwind of one of the larger mountains, and I told the controller that I'd take a VFR climb for a while to work at minimizing the turbulence. That was okay, and I was on my way up to 11,000 feet, the filed altitude.

I got up to 9,000 feet very handily, with the help of an updraft, but then wound up in a downdraft. At the best rate-of-climb speed and maximum power, the airplane just wouldn't climb at all so I asked the controller for 9,000. From the view ahead, that altitude would be comfortably on top of all clouds so why try to go higher when it wouldn't seem to work anyway?

The flight at 9,000 was one constant fuss with up- and downdrafts for about an hour and a half. For a while I'd have full power and best rate-of-climb speed, just barely clinging to 9,000. Then it would be bottom of the green on power, top of the green on airspeed, trying not to climb. There was very little turbulence, just the ups and downs.

The plot thickened when I was in the vicinity of Richmond. Cloud tops were building. A helpful controller said that another airplane had found tops at 9,500 so I tried 11,000. Then 12,000. At 12,000 I found the horns of a dilemma. Tops were still higher, I had oxygen on board but was only 130 nautical miles from the destination and didn't particularly relish going to 15,000. The temperature was -18 C, but every time I'd bust through the tops of stratocumulus, I'd get a little ice. It was too cold for ice anywhere but in the tops of stratocumulus, and those were the clouds at hand so that was the game I had to play.

I heard a Skymaster pilot flying in the opposite direction at 8,000 report some light rime ice. With the controller's permission, I asked him where he had come from and what conditions he had encountered. He had come down the route I was about to go up, and he hadn't had any ice at 8,000 until getting close to Richmond. I was passing Richmond so I took this as an indication that I could go down and find an ice-free altitude for most of the rest of my flight. The controller approved, and at 9,000 feet we soon flew into an area of nothing but lower clouds, and ice crystals up high. It was virtually clear on to Trenton, where the strong northwesterly winds offered the only diversion on landing.

In retrospect, we must have passed through a trough in the Richmond area, extending from the low over Buffalo to the low southeast of New York. With two lows so close together, I should have anticipated some weather in a trough between them. Later, the trough moved north and we had snow in the early evening in Trenton. By the next morning the lows had merged, with the one off the coast the identifiable survivor. It had moved far northward, up the east side of the trough at the 500 mb level.

In this case, the pressure near the center of the low was quite low—29.45 was the altimeter setting at Trenton when we landed—and the circulation was moderately strong. The reason we found relatively good weather was because we were on the backside of the low over New York. By the time the circulation made it around that low and came back into the Trenton area as a northwesterly wind, there just wasn't a lot of moisture left in it. It's hard, in retrospect, to see why they forecast 1,000 overcast and three miles for our arrival time but perhaps that was based on the arrival of that apparent disturbance in the

Richmond area. You often note that when there is more than one low in an area, all the systems tend to lose strong identity and fronts become more diffuse. That is, until and unless the lows merge and make one strong system.

NOVEMBER—TRENTON TO LITTLE ROCK

Stationary Front

According to the FSS, the synopsis showed a warm front running from the Virginia coast westward to southwest Arkansas where the front became a cold front and trailed down into Texas. The briefer wouldn't position a low along the front. He said that there wasn't one and that the fronts were hardly moving. I felt that the briefer had called a warm front a stationary front. At least the front wasn't moving, as a warm front will do.

The sequence reports were all okay. The lowest one along the route to Lexington, Kentucky, a planned fuel stop, was Elkins, West Virginia, at 2,000 overcast and five miles visibility. But there was a chilling addendum—freezing rain. This was reported at Elkins and Lexington. The forecasts at Lexington and along the way called for generally good ceilings and visibilities with occasional lower conditions in fog early in the day, and in thunderstorms later on. The forecaster anticipated either northward movement of the front or the formation of waves or a low pressure system along the front. Surface temperatures early in the morning were all around freezing. Winds aloft forecasts called for light southwesterly winds at 6,000, with the temperature at that level forecast above freezing by the time I reached the Charleston, West Virginia, area. The 9,000 foot winds would be in the twenties, southwesterly, with 0 C forecast at that level all the way to Lexington. After 10 A.M. the temperatures aloft were forecast to warm to above freezing at 6,000 and 9,000 with the wind velocities largely unchanged. The only mention of icing was in relation to light freezing rain in West Virginia. The freezing level information called for multiple freezing levels.

There was a high overcast at Trenton on departure. The sun was just rising, and the clouds to the west were indeed strange

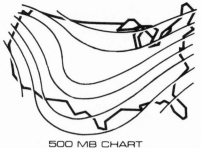

SURFACE CHART   500 MB CHART

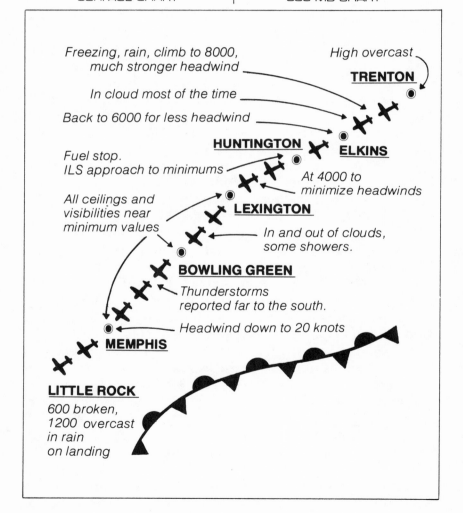

Freezing, rain, climb to 8000,
much stronger headwind ———

High overcast

TRENTON

In cloud most of the time ———

Back to 6000 for less headwind ———

HUNTINGTON   **ELKINS**

Fuel stop.
ILS approach to minimums ———

At 4000 to
minimize headwinds

All ceilings and
visibilities near
minimum values

LEXINGTON

In and out of clouds,
some showers.

BOWLING GREEN

Thunderstorms
reported far to the south.

Headwind down to 20 knots

MEMPHIS

LITTLE ROCK
600 broken,
1200 overcast
in rain
on landing

in appearance. I was level at 6,000 and there were numerous snow showers around. A station or two was reporting snow. The groundspeed at 6,000 was only 20 knots off the Cardinal RG's usual pace, so I felt rather comfortable about that. The air had a nervous jiggle to it. Not real turbulence, but a reminder that something might be brewing.

Any early morning complacency was broken when I was between Lancaster, Pa., and Martinsburg, West Virginia. The temperature at 6,000 had been right at freezing and had prompted some idle thought about whether it was just below or just above. A shower answered the question about temperature. Freezing rain. A call to Center revealed that others had reported freezing rain in the vicinity. Three alternatives: retreat; land quickly, while still in visual meteorological conditions; or climb, per the usual bromide for freezing rain.

I chose the climb. The rationalization was that I should find warmer air aloft. It wasn't forecast to be there, but the freezing rain suggested that it would be there. Also, I could always detour to the south and find warmer air. Roanoke, Virginia, had been added to the list of weather reports and the surface temperature there was +9 C.

I learned a lot about the weather of the day when passing through 7,000 feet in a climb to 8,000 feet. There was enthusiastic turbulence, the temperature rose to +5 C, and the ice slid right away. It was a rather classic picture of the slope of a warm front, or of an entry into a warmer layer of air that is overrunning colder air below. I also learned that the winds aloft forecast of the day was as inaccurate as the temperature forecast. The groundspeed at 6,000 had been 120 knots; it dropped to 88 knots at 8,000 feet. The actual wind at 6,000 had been close to the forecast, as had been the temperature. At 8,000 the wind was about 30 knots stronger than forecast and the temperature was 4 degrees above forecast. Given the freezing rain below, I'd accept the wind for the temperature increase, but it would have been nice to have a bit less breeze up there. The day would become long at 88 knots.

When the wind forecast is substantially in error, it's time to look for other mistakes in the forecast. This day the error was only in wind speed. The direction was okay. I felt that this would mean warmer than forecast temperatures, as already observed, and perhaps a mean streak in the weather much closer to the front. For now the air was smooth at 8,000 feet as the airplane droned along in the clouds.

The flight plan had been filed to Lexington, but I gave up on that destination pretty early in the flight. At 90 knots average, Lexington would take almost five and a half hours. That cuts it too thin. There couldn't be any alternate for Lexington. My flight plan, as filed, listed Cincinnati as the alternate. It was both legal and good based on the forecasts, but in the real world that became a fiction. Huntington, West Virginia, became the chosen spot in my mind, or I might even land at Charleston, West Virginia. I would decide when closer.

By nine o'clock, the plot started to thicken. I'd be passing Elkins in a few minutes; it was down to 400 broken, 2,000 overcast, five miles visibility in light snow, temperature at 0 C and dewpoint but two points less. Charleston was 400 overcast and a mile and a half. Huntington was hanging in there with 4,400 overcast, two and a half miles visibility. The weather was still relatively warm and spiffy to the south and I thought I'd keep flying toward Huntington, with the thought that a southerly diversion to Tri-Cities would solve any problem of cold air at the surface as well as offering reasonable weather for an IFR arrival. Also encouraging was word that nobody was reporting ice during arrival or departure at Charleston.

The warmer air was lower in West Virginia and a descent to 6,000 maintained a plus four on the temperature and cut the loss in groundspeed to 35 knots. That was better. The Huntington weather worsened, though, and was now 300 broken, 4,000 overcast, and one mile. Warmer air over cold ground and warm rain falling into colder air will take its toll every time. The same principle that makes steam in the shower makes low clouds, so this shouldn't have been any surprise. The approach at Huntington was in minimum conditions. The surface wind there was northeasterly, and while some wind shear effect might have been anticipated from the difference in surface and aloft winds, there was only a bit of light turbulence toward the last of the approach.

It's not normal to fly all day in the influence of one weather system, but my briefing for the continuation from Huntington on to Little Rock offered more of the same. The briefer said that the front was still stretched out south of my course, that there was no low shown on the map, and that the only low he could find was on the prog chart for the next morning. That one was forecast to be in Louisiana. At least the absence of a low was an indication that forecast thunderstorms wouldn't be numerous, unless of course a low pressure center formed earlier than anticipated.

Ceilings were uniformly low on to Little Rock and the southwest wind was to persist. The wind was forecast at 15 knots at 3,000 feet so I filed for 4,000. Thunderstorms were forecast all along the route but none were showing on the latest radar summary chart. It's 540 nautical miles from Huntington to Little Rock. I didn't want to file all the way because there was no gold-plated alternate, and so I filed to Memphis with Little Rock as the alternate. The situation could be reappraised later.

Flying at 4,000 in cloud some of the time and between layers at other times, with occasional periods of showers, it was obvious that there was something more going on than just a front south of my course. The periods of showers alternating with between-layers conditions suggested that weak waves were probably moving northeastward along the front. And some thunderstorm activity farther to the south suggested that, closer to the front, the activity was getting enough support to build into something substantial. I did find the wind aloft to be more southerly than southwesterly.

All the while, the surface winds remained northeasterly. It was a huge area of a very shallow and weak northeasterly flow overrun by a southerly flow. The temperature differences became more dramatic, too. It was +3 C on the surface at Little Rock and about 10 degrees warmer at 4,000 feet.

Approaching Memphis, a continuation on to Little Rock looked good so I kept on going. The Little Rock weather was 600 broken, 1,200 overcast, and four miles on arrival with a northeast wind and a cold rain. There was a third leg to the flight, on to Wichita, Kansas, and it was flown in steadily improving conditions. In fact, it was sparkling clear at Wichita— a nice respite after all the clouds and rain.

These fronts that languish across the Appalachian mountains and down into the Gulf of Mexico bear watching. No low formed on this front this day, but if a strong one had developed and moved to the northeast, then the weather along my route could have been wetter and much more turbulent as the flight path moved through or a bit north of the low.

Why didn't a low form this day? Probably because most of the action was in the lower levels. The surface flows on each side of the front were rather weak and the frontal zone was rather diffuse. There was little upper level support, too. It was mainly a lot of warm air moving up over a shallow layer of cold air at the surface. The system was slow in being affected, but the developing trough on the 500 mb chart, with the tip in central Mexico this day, later spawned the development of a strong low on the front that moved almost straight north, under the east flank of the trough. It was in northern Michigan a couple of days later.

DECEMBER—WICHITA TO TRENTON

Two Pair

According to the briefer, the surface map showed two cold fronts and two warm fronts, all stemming from a low pressure center in northern Ohio. Looking east from Wichita in the early morning hours of this first day of December, the initial cold front was in central Illinois; the second was in central Ohio. A warm front extended southeastward from the low out into the Atlantic off the coast of North Carolina and a second warm front extended east from the Ohio low and into another low just off the Jersey coast. It sounded like a complicated situation and indeed the briefer characterized it as a "complex" low pressure system. One thing to consider in such a situation: Nature can't play in spades in that many fronts. One might be mean and the rest relatively benign. Or perhaps they are all benign. The challenge is to find out what's what.

The weather was clear in Wichita. Moving to the east, St. Louis and Indianapolis were 700 overcast; the visibility was good at St. Louis and restricted in snow at Indianapolis. The Indy temperature was +4 C. Not far to the east, Dayton was 2,000 overcast and twelve miles, +12 C on the temperature, with southwesterly surface winds gusting to 32 knots. The winds aloft were forecast to be northwesterly over Kansas,

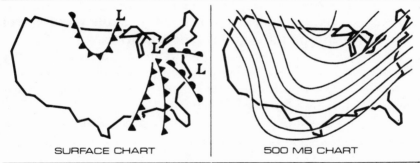

SURFACE CHART   500 MB CHART

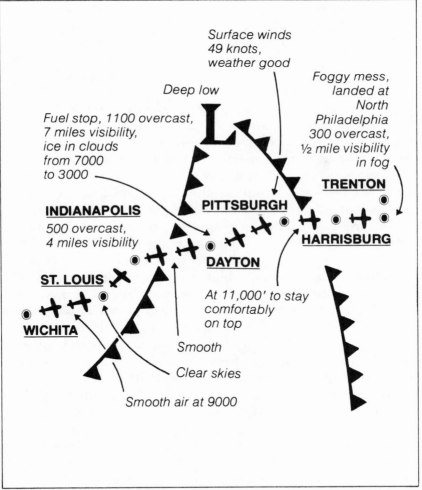

Surface winds
49 knots,
weather good

Deep low

*Foggy mess,
landed at
North
Philadelphia
300 overcast,
½ mile visibility
in fog*

*Fuel stop, 1100 overcast,
7 miles visibility,
ice in clouds
from 7000
to 3000*

TRENTON

INDIANAPOLIS

PITTSBURGH

*500 overcast,
4 miles visibility*

HARRISBURG

DAYTON

ST. LOUIS

*At 11,000' to stay
comfortably
on top*

WICHITA

Smooth

Clear skies

Smooth air at 9000

Flights: Fourth Quarter/October, November, December / 229

swinging to westerly over Missouri and Illinois and southwesterly over Ohio in a classic circulation around the south side of
a low. Velocities in the twenties to begin, increasing to 35 knots
at 9,000 feet in Indiana. The Dayton and Columbus forecasts
were okay for IFR, but with strong surface winds on the menu.
The briefer had no Sigmets or Airmets, but we agreed that it
would probably be bumpy down low.

Aloft, I learned quickly that the wind forecast was in error in
direction and velocity over Kansas. Instead of northwesterly at
25 it was straight westerly at 40 knots.

An hour into the flight, I checked weather and found that St.
Louis had cleared. All other stations had an overcast at around
1,000 feet, plus or minus a few hundred, with good visibilities.
Tops at Indianapolis were reported at 5,000. Temperatures
were dropping, though all stations were still above freezing.
The Columbus, Ohio, surface wind was gusting to 43 knots.
Lots of action down there but very smooth at 9,000 feet. As I
progressed toward Dayton, all stations were okay with the exception of Indianapolis which dropped to 500 overcast and
four miles and stayed there for a couple of hours.

Top reports varied from 7,000 to 9,000 feet. I was cruising
at 9,000, well on top approaching Indianapolis, but I could see
some higher clouds ahead. The temperature at 9,000 was
below freezing, so I wasn't eager to get into many clouds, but
these turned out to be very thin and wispy, offering nothing
more than an ever so gentle frost right at the leading edge. As
I passed Indianapolis, the higher clouds abruptly ended. The
lower layer that I had previously seen, with tops at 5,000, was
now higher, probably around 7,000 I guessed. The little area
could have been the most westward of the cold fronts.

Dayton was reporting 1,100 overcast, seven on the visibility,
temperature +7 C, wind from the southwest at 18 knots with
gusts to 26 as I started a letdown to land for fuel.

The cloud tops were at 7,000 feet, and I was rather surprised
at the turbulence and considerable ice accumulation I got when
I was level at 7,000 awaiting further descent. The top of a
stratocumulus will do it every time, though. The accumulation
continued when I was at 5,000 and 3,000. The temperature
went above freezing at 2,500 and the ice began to slide off as
I started down the glideslope. I was in the clouds and below

freezing temperatures for twenty-two minutes and accumulated about a half inch of very rough ice. It was enough to cost from 15 to 20 knots of airspeed.

There was quite a bit of turbulence on final, and the wind was still strong when I parked the airplane and went in to check the weather for a final leg to Trenton.

The FSS started the briefing with real horror stories about Sigmets for severe turbulence, Airmets for icing, and other assorted bad things. One cold front was ahead of me, the other was behind. The briefer didn't mention a warm front. Pittsburgh was in good shape—2,000 broken and ten—with very strong southwesterly winds, gusting to 49 knots as a matter of fact. All weather east of the Appalachian mountains was down around 500 feet, with North Philadelphia forecast to be 1,400 overcast and three miles at the time of arrival. Tops, he said, were reported up to 11,500 feet. The winds at 9,000 were forecast to be westerly at 50 knots over Pittsburgh (but one knot better than the surface wind), slacking to westerly at 40 knots farther east. There were no pilot reports of turbulence.

The Sigmets about turbulence seemed to me the most significant weather. I knew there was ice in the clouds, but from the amount I accumulated on the letdown in twenty-two minutes, I figured that I'd not get enough to worry about in a ten minute climb at low airspeed. How to handle the turbulence? Have a look. It probably wouldn't extend too high and I had a full oxygen bottle on board.

Enough ice accumulated on the climb to cost about 8 to 10 knots, but that didn't matter because of such a fine tailwind. The Cardinal RG's true airspeed was 130, the groundspeed was 180. Couldn't complain about that. The cloud tops looked flat as a table ahead, and the controller said that he hadn't had any reports of turbulence. Airplanes were moving about, even in and out of airports ahead and in rough country. They had one on an approach to Latrobe, Pa., where the surface wind was gusting to 45 knots.

As they will do, the cloud tops sloped upward over the mountains. That most easterly cold front was in fact over the mountains. Still, it appeared that 11,000 would clear the clouds comfortably so I moved up to that altitude. Then, just west of Harrisburg, Pa., the undercast ended and all I could see was what appeared to be a blanket of fog from the mountains eastward to infinity.

I was relishing the fact that the groundspeed had moved to 190 knots and that I'd be home by 4:30. Everything looked good. Checking weather en route, the latest I had marked down showed North Philly with 500 overcast and a mile and a half visibility. That was well below forecast but okay. The last FSS I had spoken with didn't have the Trenton weather.

Then the bad news. The Center controller called, said to hold at Lancaster, and added that Trenton was reporting 200 overcast and a half mile. Then a minute later he called back and said that Trenton was now zero-zero, North Philadelphia was 300 overcast and a half mile, and Philadelphia International was open only intermittently. Traffic was backing up.

I changed over to FSS while I held and found that Newark, Morristown, and Allentown were all below ILS minimums. The haven became Baltimore. I could go there if necessary. I had plenty of fuel to hold a while, go on to Trenton and try an approach there or at nearby North Philadelphia, and then move to Baltimore.

The clearances came slowly and I moved from one holding fix to another and finally toward Trenton. A quick check with the tower revealed that they were still zero-zero so I shot an ILS at North Philadelphia which was still reporting 300 overcast and a half mile. At least I was close to home.

I didn't think about it until after I landed, but I hadn't hit a bump since breaking out on top at Dayton. There was not even any shear on the approach at North Philadelphia. The surface wind there was light southwesterly; aloft it was strong southwesterly but the transition must have been relatively gradual. The surface temperature at Philly was +10 C—it was almost that warm at 11,000 feet over Harrisburg.

The widespread fog could be explained by the strong moist warmer flow over cooler ground—it had been quite cold for a while before this day. When the flight started, the major consideration was turbulence related to strong winds. The major problem turned out to be fog. That just shows once again that you must take weather as it comes, not as it is forecast.

The weather question this day related to how such an active map could have had all the clouds (and all the action) in altitudes below 7,000 feet except for a short stretch over the mountains where some clouds extended up to 9,500 feet. Looking at a map with two cold and two warm fronts to fly through just doesn't suggest a smooth ride on top of all clouds in a low altitude airplane. And when there is a westerly surface wind of almost 50 knots blowing on the west slopes of the Appalachians, there is usually turbulence up to well above 10,000 feet.

The key this day was in stability, as indicated by temperatures aloft. Over Dayton, the temperature at 9,000 feet was −2 C, on the surface it was +3 C. That's only 5 degrees colder from the surface to 9,000. Along the east coast there was a slight inversion at about 3,000 feet, where the cloud tops were located, and the temperature was all but steady from there on up to 11,000 feet. The wind patterns were moving warm air over cool air; up higher, at the 18,000 foot level, there was no great influx of cold air running over the warm air below so there were no vertical developments of anything. I never really felt like there was an identifiable warm front, either.

Too, the low was well north of my course. The FSS said it was in northern Ohio but the real chart showed that the low never was there. It was in northwestern Indiana at 1 A.M. and had moved to northern Michigan by 7 A.M. The en route clue I had to the FSS's misplacement of the low came after takeoff at Wichita when the wind aloft was found to be westerly and strong instead of northwesterly and moderate. That suggested that the low would be north of my course.

DECEMBER—TRENTON TO ROCKY MOUNT TO SAVANNAH

Front or Trough?

The "Today" map depicted no frontal activity to the south, only a cold front to the west plus some rain over Virginia, the Carolinas, and Georgia. It did include mention of isolated severe thunderstorms in South Carolina and Georgia. The FSS briefer said that there was a stationary front just south of Trenton, but no other prominent map features toward Savannah. Trenton was reporting 200 overcast and three-fourths of a mile

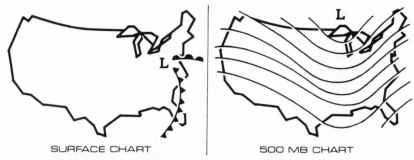

SURFACE CHART   500 MB CHART

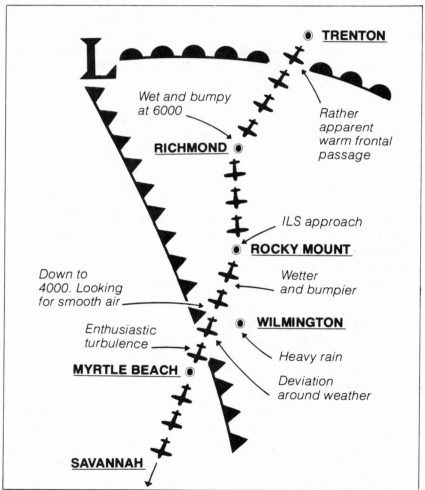

L

○ **TRENTON**

Wet and bumpy
at 6000

Rather
apparent
warm frontal
passage

RICHMOND ○

ILS approach

ROCKY MOUNT ○

Wetter
and bumpier

Down to
4000. Looking
for smooth air

WILMINGTON ○

Enthusiastic
turbulence

Heavy rain

MYRTLE BEACH ○

Deviation
around weather

SAVANNAH

visibility; all other reporting stations save Savannah were reporting good VFR conditions. The radar summary showed scattered thunderstorms along the coast and inland a couple of hundred miles; the wind forecast for 6,000 and 9,000 feet projected a 33 knot headwind. It was cold on the surface but ice wouldn't be a problem because of warm overrunning. Because of the headwind, and because of forecasts of marginal alternate weather, a non-stop to Savannah wouldn't be possible. I filed to Rocky Mount, N. C., as a fuel stop.

The temperature increased from +4 C at the surface to +15 C at 4,000 feet as we climbed away from Trenton; then it dropped back to +7 C at 6,000 feet. Lots of instability started at four. I could feel it as the airplane lurched through turbulent rainshowers. After about thirty minutes, though, it smoothed out a bit. Apparently I had flown through that stationary front the briefer mentioned.

The 86 knot groundspeed was quite bothersome. That was 54 knots under the true airspeed, and reflected a much stronger flow than had been forecast. What did that mean in relation to the surface map that I had in mind, and to the terminal forecasts written on my flight planning sheet? The low was deeper than forecast, or perhaps there was one there that didn't make it to the map described in the briefing. Or could the front be farther east than depicted?

The en route weather was forecast to be pretty good as far as Richmond, but all these forecasts started falling apart. Instead of good VFR, stations started reporting zero-zero in the Baltimore and Washington area. Clouds and rain became prevalent at 6,000 feet. At one point early in the flight I had actually considered trying VFR at low level, to get out of the wind and the bumps, but it was fortunate that I didn't give that more than a passing thought. As wet and bumpy as it was IFR at 6,000, at least I didn't have to worry about staying clear of clouds.

There were periods of smooth air and between-layers flight, and there were what seemed to be lines of rainshowers with light to moderate turbulence as I proceeded very slowly southward. The groundspeed varied between 81 and 90 knots. The Stormscope did not depict any electrical activity to the south, and the only pilot reports available from the controller indicated light to moderate turbulence and rain along the route.

Conditions remained much the same to Rocky Mount. There the report was 3,500 overcast and two miles visibility, but it was actually right at minimums for my approach. Looking straight up after landing it appeared they did have a 3,500 foot ceiling, but from above, the obscurement looked more like a very low and very solid overcast. I made an ILS approach; another airplane tried a VOR approach and had to come around and shoot the ILS to get down.

A visit to the FSS on the field at Rocky Mount gave me a chance to study the weather charts, and the depiction was far different than that I had gotten from the FSS at home and from the "Today" map. There was for a fact a warm front south of Trenton. I had flown through that and it was history. But there were three lows to the west, one in the Gulf, one in Kentucky, and one over the Great Lakes. These were joined by a cold front. A north-south trough line was depicted over the central portion of Virginia and the Carolinas, just west of my route. The wind at 6,000 feet was forecast to be southwesterly at 29 knots but I knew that it was stronger than that. There had to be some correlation between the stronger than forecast headwind and that trough.

Stations toward Savannah were all reporting poor weather, but nobody had a thunderstorm. The radar summary chart was still relatively good, with only scattered activity shown.

Back up at 6,000 feet, I found the going still wet and bumpy. Wilmington was reporting heavy rain when I passed there, but in response to my query about cells, they said that they didn't seem to be painting what appeared to be a cell. It was just one big area of heavy rain. The ride I was getting verified that the downpours were predominantly rainshowers. There was very little up and downdraft activity in them.

When I was close to Myrtle Beach, the situation changed markedly. I was at a lower altitude, 4,000, sought in hopes of finding smoother air, and flew into a between-layers situation that was quite unique in appearance. The lower clouds had little wisps extending upward—they looked like candy kisses—and the higher clouds had a rather dark and strange appearance. Ahead, there were build-ups that looked for all the world like thunderstorms. But the Stormscope showed no electrical activity and the controller said there were no identifiable cells.

The turbulence in this area was particularly bothersome. There weren't any up- or downdrafts, rather it was the sharp jab type that seems to grab the airplane and make it shake and shudder. The clouds that looked like storms were oriented north-south and as I got close to them the turbulence increased in intensity—almost to the point that the airplane was difficult to control. I was flying by eyeball at this time, a good distance away from the clouds, but the deviation heading selected had a basic flaw. It was taking me out over the ocean.

In quizzing the controller, I found that there was now a fairly intense weather return on his scope off to my right. I told him that I'd like to fly a heading of 180 until it looked better to the right. He okayed that, and a few minutes later said that a heading of 250 would run through a light spot.

It was quite bumpy for a minute, including one downdraft that caused a loss of 600 feet before I could arrest the rate of descent, but then things started smoothing out and I was soon between layers. The trough line I saw on the map at Rocky Mount must have turned into a true surface front, and I had just passed through that front. The surface winds moved around to westerly (from southeasterly), the weather beneath started improving, and the trip on to Savannah was uneventful.

In composing a pilot report, I was mulling whether to call the turbulence moderate or severe when I noted that my Jeppesen book was still perched serenely on the right seat. It had never moved. And then I recalled that in the course of all the activity I hadn't once moved against the safety belt. Given the reporting criteria for turbulence, that meant it qualified only as light turbulence. I had to report it as moderate.

In retrospect, I did a poor job of staking out this day's weather. I kept thinking that I was flying in the southwesterly flow of a warm sector, a lot of miles away from the low pressure center and the cold front. Actually, I was flying parallel and very close to an active cold front up to Myrtle Beach, where I passed through that cold front. After getting through it, I got the message about the real synopsis. It wasn't as depicted on the charts and I should have figured it out sooner—when you have that much southerly flow and instability over that great a distance, you almost have to be close to and paralleling a front.

Epilogue

One of the nicest things about flying and weather, and flying in weather, is that there is always something new to learn. The other things that we do with airplanes might reach a plateau, where there aren't a lot of fresh things to do or see, but weather just isn't like that. The elements are dynamic. No two low pressure systems seem to be alike and fronts come in many varieties. The challenge is without a limit.

So let me stress that the flights related in this book don't cover all the situations that might be encountered. But they do provide a good representative selection.

The exercise of recording briefings and comparing them with what was actually encountered, and with the official National Weather Service maps published later, was enlightening. I found a greater variation than I had thought between the available information from the flight service station and the real world. This was all compiled on a random basis; no effort was made to compare only the difficult situations. What is here is simply what occurred on flights with enough weather to mention during a one year period.

It is quite clear that a pilot who does not devote time to the study of meteorology is either in for a lot of weather-related misadventures, or a lot of wasted time pacing the floor at the airport wondering whether or not it is okay to fly. To operate in anything other than perfect weather conditions, a person just has to have his own weather interpretive ability. It is not required for any license or rating, but it is one of the most important things in flying.

Becoming a weather-wise pilot is fun, too. Much of the mystery disappears. Frustrations vanish when you *understand* the situation that stands in the way. When you cancel a flight, there's not that gnawing sense of doubt. It was cancelled because *you* decided not to go, based on available information plus the principles of meteorology. When you decide to go, it is with more than cursory information; it is with that information plus knowledge of the subject. You can't do any better than that.

With an understanding of weather, the things we see and feel take on new meaning. Situations that used to be illogical become quite reasonable. There is still the urgent requirement to gather all available information while en route, but if you are ahead of the situation, the information will bring verification more often than it brings surprises. When it does bring a surprise, you can use it in teaching yourself the next lesson—something that you should continue to do for as long as you fly.

Index